MUNI
PRESS

MASARYK
UNIVERSITY
MONOGRAPHS
Vol. 1

HANA LIBROVÁ
and Vojtěch Pelikán – Lucie Galčanová Batista – Lukáš Kala

THE FAITHFUL
AND
THE REASONABLE

CHAPTERS ON ECOLOGICAL FOOLISHNESS

Illustrated by Bohdan Lacina
Translated by Graeme Dibble

Masaryk University Press
Brno 2021

This book was supported by the "Scientia est Potentia" Masaryk University fund. The Czech Edition was made possible thanks to the financial support of the Czech Science Foundation as part of the project "Environmentally-Friendly Lifestyle in the March of Time" (GA15-05552S).
The publisher would like to thank the creators of the illustrations, photographs and poems for their permission to use their work in this book.

Book reviewed by:
Prof. RNDr. Bedřich Moldan, CSc., d. h. c.
Prof. PhDr. Dušan Lužný, Dr.

Pandora's box is open.
Not my children or grandchildren but I myself look on
as the present day redraws the map of the world
with rusty needles in the tops of the pines
and green scum on the surfaces of ponds
from poisonous algae. Suddenly there is nowhere
to escape to across the water in my bark canoe.
All that remains of the creek is a wadi
better suited to a caravan of Tuareg nomads.
Come August the maples are stained with the blood
of punctured lungs, and entire nations
set out in search of a new home,
while other nations pin their hopes
on the night sky, searching for a new Earth
hundreds of light years away
and just as unattainable as the place
where we were once born.

Radek Štěpánek (2018, p. 36)

Contents

Foreword

The book you are holding is the third part of a trilogy which deals with changes in environmentally friendly lifestyle, as captured by three field studies between 1992 and 2015. In the first volume, *Pestří a zelení: Kapitoly o dobrovolné skromnosti* (The Colourful and the Green: Chapters on voluntary modesty) from 1994, I tried to answer the question of whether in the early 1990s, at a time characterized by a general escalation in consumption, there was a way of life in the Czech Republic that could be termed *voluntary modesty*. If so, what form did this life take? Ten years on, I went back to the surveyed households which I labelled – due to their diverse social backgrounds and various motivations – "the Colourful". I held the hypothesis that a modest way of life contains some features of *ecological luxury*.[1] In 2003, I published a theoretical analysis of this phenomenon, historical analogies and research results in the second volume, *Vlažní a váhaví: Kapitoly o ekologickém luxusu* (The Half-Hearted and the Hesitant: Chapters on ecological luxury) I examined the question of what role ecological luxury might play in the spread of an environmentally friendly lifestyle envisaged by some authors (Elgin 1981; Grigsby 2004; Alexander 2015). This third volume considers the lifestyle through a variety of theoretical approaches, from the perspective of ontology, psychological attitudes, individual motivations and the religionist point of view.

However, the main focal point of the book remains the empirical research into families. Over nearly a quarter of a century that has elapsed since the first research contact, I have established a friendly relationship with my respondents. Naturally, the rules of sociological research require researchers to reflect this fact and take it into account in their interpretations.

The content of the book is founded on one other basic stance on the part of the author: Unlike most (if not all) sociological works on environmental issues, my primary concern is the prosperity of nature. I cannot express this allegiance better than in the words of my colleague, the philosopher Václav Bělohradský: "Being on the side of the butterflies is a categorical imperative of our time."[2]

The preparation of this sad book entailed a great deal of joy.

* There was the joy of working in the research team as Vojtěch, Lucie, Lukáš and I thought about the Colourful families, discussed the conception of the research and argued about how to interpret it after returning from the field.
* For me, it was a great joy to revisit the Colourful families as my research career drew to a close. I wouldn't have managed it on my own, but I was accompanied by young colleagues who took charge of organizing the fieldwork and filled in for me when necessary.

* In their research into the children of the Colourful, my colleagues experienced the joy and excitement of discovering something new. Through their research, they have also stored up another joy for themselves in the future, as their audio and visual records will become even more valuable over time.

* During moments of joy in investigating the modest way of life, I was sorry not to be able to share them with Josef Vavroušek, the man who instigated all of this. If he had heard our interviews with the Colourful families, it would have temporarily restored his faith in the existence of the Third Way.

* After the joy of research came the joy of writing, accompanied by worries and uncertainty. This joy wasn't even diminished by the challenge of digitally processing the final form of the manuscript since we were guided by the sure hand of Michaela Kašperová.

* In order to find out if the text was at all readable, I subjected the manuscript to the criticism of the students in my seminar group. They were kind, critical and inspiring.

* As I write this, my thoughts turn to someone who is no longer with us, Tomáš Ryšavý. He was a typical organizer of small-town cultural life and one of the sceptics among the Colourful. He would have been sure to defend me against any readers criticizing the chapter on environmental grief.

* I found a paradoxical joy in the poetic depiction of environmental grief in Radek Štěpánek's verses. Will it be shared by the readers?

Should I, at the close of this foreword, paraphrase and modify the opening words to the introduction: 'Creatures are dying, but we have our own joys'? There might not be enough of that joy to counterbalance the sorrow of reality, the sadness that arises from environmental awareness and reflection. In the end, though, readers can judge for themselves.

Hana Librová, August 2016

---

[1] Ecological luxury is environmentally friendly luxurious behaviour which, unlike predatory luxury, is capable of self-restraint; consciously or intuitively, it reduces the ecological footprint (Librová 2003, p. 61).

[2] Bělohradský made this statement on Czech Television in the programme *Nedej se!* (Stay Strong!) in February 2002.

I.

Introduction: Creatures are dying, but we have our own worries

Is it the fault of journalists? To scare or not to scare?

In the autumn of 2014, the World Wide Fund for Nature published frightening information about the state of life on the planet (McLellan *et al.* 2014); it claimed that since 1970 the number of mammals, birds, reptiles, amphibians and fish had fallen by 52 percent.[3] Clive Spash, editor-in-chief of the major scientific journal *Environmental Values*, responded in an expressive way; he wrote in blunt terms that the Living Planet Index is a measure of death rather than life (2015). Even more dramatic are the reports of the extinction of invertebrates such as insects (Čížek *et al.* 2009). These alarming messages fail to elicit an appropriate response from the public. Can we blame mere mortals for focusing on their own joys, worries and interests, for not sharing the consternation of the biologists? As well as being caught up in our personal concerns, we hear disturbing news about goings-on in the world. These days we can hardly expect adverse developments in society to spark an interest in nature.[4]

Even among my own colleagues, I don't know anyone who would lose sleep over the drastic decrease in blister beetles or the collapse of bluethroat populations. Only occasionally do I hear them express concern about the persistent drought in South Moravia. I hardly ever hear people on the street lamenting the disappearance of sparrows, and Vladimir Putin's futile attempts to save the Siberian tiger only appear on the internet as a jokey item from the Kremlin.

The journalist George Monbiot describes "almost certainly the greatest environmental disaster of the 21st century – so far": "Fire is raging across the 5,000km length of Indonesia. (...) The air has turned ochre: visibility in some cities has been reduced to 30 metres. Children are being prepared for evacuation in warships; already some have choked to death. Species are going up in smoke at an untold rate. (...) It is surely, on any objective assessment, more important than anything else taking place today." Monbiot asks: "So why is the world looking away?" (2015b).

When the philosopher Karl Jaspers writes "blindness for the misfortune of others, lack of imagination of the heart, inner indifference toward the witnessed evil – that is moral guilt" (1965, p. 64), he is only thinking of the offences we inadvertently commit against our fellow human beings. That's how we were taught, and so I'm not angry or even surprised that people don't get worked up about the drastic deprivation and impoverishment of nature. But how is it that they don't even think about the fact that the state of nature affects and will affect their lives and the lives of their children?

Is the public's lack of interest in the global state of the environment the fault of the media, which do not provide enough up-to-date information about environmental developments? The fact is that newspaper reporters only take a one-off interest in spectacular conflicts or, perhaps, in the sensational return of large carnivores to the Czech countryside. Only occasionally is the reader presented with informed analytical commentary about complex ecological issues. It's a vicious circle: The editors of the press, television and radio broadcasts state in their defence that they just write/talk about what their readers/listeners are interested in; environmental questions are not among those topics.[5] The media thus reproduce and reinforce the stereotypical appetites of readers and, willingly or unwillingly, serve the power structures that are generally inimical to environmental efforts. Clive Hamilton points to the sociological and political dimension of concealing unpalatable environmental truths, specifically climate disruption. The fossil fuel lobby consciously lies about the environmental impact of its actions and uses the media to deliberately "throw sand in the eyes of the public" (Hamilton 2014a, p. 335).

By contrast, other authors believe it is almost impossible to avoid information about climate change (Whitmarsh 2011). Jon Krosnick and his colleagues from Stanford University have shown that people lose interest in climate change when the sheer mass of information makes them aware of the size and complexity of the problem and the extreme difficulty of solving it (Krosnick *et al.* 2006, p. 34).[6] Some psychologists, environmental educators and activists recommend communicating negative information to the public in small doses[7] and in an easy-to-swallow format, if at all. Hamilton and Kasser aptly characterize the softly-softly approach taken by most environmental organizations: "Don't scare the horses" (2009, p. 8).

Fearfulness or timidity is probably something held in disdain by all cultures. The cornerstones of European culture are clearly laid down: Moses's assurance as he led the people out of Egypt and the basic message of the Gospels are: "Do not fear!" However, Hans Jonas, a philosopher who deals with the ethics of nature in a technological civilization, draws attention to the special case represented by fear and the role of fear in the emergence of ecological responsibility. The long-term consequences of human activities cannot always be gauged using scientific methods – it is also necessary to mobilize our sensitivity and intuition.[8] Fear "has a bad moral and psychological reputation in better society, (...) we need to speak of it again, since it is more necessary

---

[5] In the Czech Republic, civic apathy is often explained as a hangover from the totalitarian regime, which destroyed interest in all public matters. When it comes to environmental questions, the passivity and indifference of Czech citizens, which also manifests itself in annoyance with activists, is evident (Librová 2014b).

[6] Here we touch upon what are known as defences of the ego, which are discussed in Chapter II.3.

[7] This seemingly harmless dilution of information can easily lead to the downplaying of environmental problems.

[8] The leading climatologist James Hansen (2011) warns of the dangers posed by inappropriate scientific reticence.

[9] This is similar to the anthropological view of Bernard Verbeek, who warns against "rose-tinted glasses". Fearlessness and the courage to take risks used to be an evolutionary advantage, but in a time of environmental crisis it is a dangerous hazard (Verbeek 1998). On the other hand, it is easy to find authors who warn against fear. Jonas explicitly takes issue with the view of Ernst Bloch, Jean-Paul Sartre and Thomas Hobbes (Jonas 1997, p. 317). For example, the sociobiologist Konrad Lorenz considers "anxious haste and hasty fear" to be one of the "deadly sins" of civilization (1974, p. 27). Sociologists often cite the principle of a "self-fulfilling prophecy" (Merton 1968). In the universal key I will mention Andrew Dobson's concept of responsible behaviour, to which an individual mustn't be motivated by reward or by fear.

[10] This fact was pointed out in the classic work The Limits to Growth (Meadows *et. al.* 1972).

today than at many other times when trust in the proper functioning of human affairs meant that it was possible to regard it as a weakness of the small-minded and faint-hearted," wrote Jonas in 1979 in *The Imperative of Responsibility* [as in Jonas 1997, p. 316; the translation given here is based on the Czech edition, which is substantially different to the English one – translator's note]. According to Jonas, fear is now a moral and intellectual virtue. "So whoever feels these sources of 'fear and trembling' (...) are not sufficiently lofty for the status of man cannot be entrusted with our fate" (p. 317). "It is also necessary to relearn respect and dread so that they can protect us from the errant paths of our power" (p. 318).[9]

In any case, what journalists offer should be enough to frantically search for an answer to the question of what to do about the current environmental situation. However, we are a long way from that position. The main cause does not seem to be the amount or nature of the information offered, but the willingness and ability to accept it.

Evolutionary deafness and evolutionary opportunity

Our deafness to negative environmental information can be interpreted in various ways – for example, sociologically, historically or politically. I am inclined towards the interpretation of evolutionary psychology and anthropology. Unfortunately, this highlights the depth and rigidity of the problem: The exhortation to think in an ecosystemic and global way with consideration for future generations is being addressed to creatures that are not evolutionarily equipped to receive it (Walletschek and Graw 1990). Our psyche shares some features with higher mammals, especially primates. Even the traits which are specifically human are evolutionarily ancient, and therefore rigid – 90% of the time that *Homo sapiens* has existed on Earth was spent as a hunter-gatherer.

The psyche of our evolutionary ancestors was guided by sensory exploration and sensory pleasure. It was oriented towards the individual's day-to-day survival, towards a small group upon which their existence depended and a space limited by walking distance. Even today, we routinely make decisions based primarily on information that relates to the near future,[10] to a small circle of people, particularly family, and to a nearby location. The extinction of species, the felling of tropical rainforests or the plight of people affected by a catastrophic drought somewhere in South America – all of this remains outside our spontaneous attention. For the behaviour of modern humans, just like their ancestors – the prehistoric hunter-gatherers – what is important is primarily simple information about rapid and conspicuous changes which can be captured directly by the five senses. Confusing mediated information about complex surreptitious processes – which undoubtedly applies to environmental ones – fails to rouse the human brain.

However, evolution also shaped the human relationship to nature in another way. Like other creatures, it endowed man with *biophilia*, an attraction to nature and the need to connect with it (Wilson 1984).[11] Human biophilia, based on our five feeble senses, is certainly weak in comparison with the biophilic instinct of other species. Nevertheless, the loss of natural senses has been compensated for by the cultivation of the human mind and capacity for rational thinking. For thousands of years, the most important source for the transformation of the biophilic instinct into a love of nature was the celebration of natural beauty through art. However, it would be ungrateful not to mention also the efforts of modern environmental education and the dedicated work of science teachers, but also teachers of humanities subjects, leaders of science and conservation clubs, Scout leaders, as well as the inspiration from environmentally sensitive parents and grandparents.

Everyday experience allows us to extract environmental information about the place where we live through direct observation. But thanks to environmental education and trained sensitivity, we can actually use the perception of a specific place to form a picture of distant or even global situations. Some of the descendants of prehistoric hunter-gatherers and heirs to the culture of past centuries are capable of using sensitive observation to obtain information about natural processes as difficult to grasp as climate change. The famous journalist Naomi Klein (2014) writes powerfully and persuasively: "The terrain on which the changes are taking place is intensely local: an early blooming of a particular flower, an unusually thin layer of ice on a lake, the late arrival of a migratory bird. Noticing those kinds of subtle changes requires an intimate connection to a specific ecosystem.[12] That kind of communion happens only when we know a place deeply."[13]

However, situations in which we can capitalize on this valuable evolutionary and cultural disposition are decreasing dramatically. What actually happens is that the sorry state of nature ceases to arouse interest. Naomi Klein notes that opportunities to identify major environmental problems through the observation of a specific site are also disappearing as a result of urbanization and spatial mobility:

[11] Wilson cites the appeal of water, which is noticeable with children, as an example of biophilia in everyday life. Apparently, each year more people in the United States visit zoos than all sporting events put together. At the same time, Wilson points out that biophilia does not always signify a positive relationship. It encompasses any kind of attachment to nature, including aggression, *e.g.* children's tendency to torment living creatures.

[12] Few people have noticed how the number of corvids – magpies, crows and jays – has risen in our countryside and cities. And hardly anybody knows what their presence means in ethological and ecological terms.

[13] *"The sun is shining with all its December might, / but the earth still remains beyond its reach. / While the treetops are already thawing, / in the umbels of ground elder, sermountain and wild carrot the hoarfrost clings on / to the silver threads of an Indian summer. / This is the third year we've set out for mistletoe along a muddy path / in places where we used to sink down into the snow,"* writes Radek Štěpánek in the poem *Christmas Day* (2018, p. 21).

"We tend to abandon our homes lightly – for a new job, a new school, a new love. And as we do so, we are severed from whatever knowledge of place we managed to accumulate at the previous stop (...) Even for those of us who manage to stay put, our daily existence can be disconnected from the physical places where we live. Shielded from the elements as we are in our climate-controlled homes, workplaces and cars, the changes unfolding in the natural world easily pass us by."

It is a tragic vicious circle: Even as we fail to notice climate change, it "is busily adding to the ranks of the rootless every day, as natural disasters, failed crops, starving livestock and climate-fueled ethnic conflicts force yet more people to leave their ancestral homes. And with every human migration, more crucial connections to specific places are lost, leaving yet fewer people to listen closely to the land" (Klein 2014).

Modern 'progress', which is already devastating nature through human demands and economic operations, is damaging it even further by weakening man's relationship to nature as a biophilic source of pleasure and the object of a love which was developed by human culture over thousands of years. It is strengthening the primitive part of our evolutionary and cultural disposition which is selfishly oriented towards the individual's day-to-day survival and the prosperity of their own offspring.

One thing should be added to the discourse on the evolutionary dimension: Let us not confuse the problem of evolutionary deafness to the damage done to nature with the question of general human selfishness, which Neo-Darwinists are convinced of (to cite just one example, Ridley 1996). Nor is the claim of their opponents about the altruistic and cooperative nature of man relevant to our problem. Indeed, the ability of people to work together, which George Monbiot clings to as a drowning man will clutch at straws (2015a), may actually further strengthen the ecological risk posed by *Homo sapiens*.

What is the book about?

Just as in the two previous volumes from 1994 and 2003, it is about environmentally friendly ways of life. However, on this occasion I expanded the set of research subjects beyond modest households. Readers will be introduced to the so-called ecopragmatists (Chapter III.2.), who regard themselves as *reasonable*. They do not cling to some form of 'untouched' nature; apparently it is necessary to reconcile oneself to its demise. It is rational to be devoted to creating a new (semi-)manmade world that will be good for people. However, the relationship to vanishing nature can also be fundamentally different; it can take the form of environmental grief (Chapter III.1.). Does this mean that anyone who is grieving, who is *faithful*, isn't reasonable? The fortunate ones are able to transform grief into the protection of living things in the field (Chapter III.3.).

The *faithful* are not just to be sought among field conservationists. Faithfulness to nature can be contained within numerous variants of everyday lifestyles. We will briefly consider life in communities (Chapter III.5.), but the main protagonists of this book will be the individuals and families who we have had the opportunity to observe in the three stages of the aforementioned empirical longitudinal research (chapters IV.1., IV.2., IV.3.). We decided, in conjunction with the publisher, to reprint the empirical findings from 1992 and 2002 with only essential modifications. We anticipate

The main focal point of the book is empirical research.

that this approach will make it easier for the reader to follow the development of the environmentally friendly lifestyle over the course of time as well as the development of the author's point of view over a quarter of a century.

Chapter III.4. will come as a surprise to most readers. It will give them an insight into a subculture which holds that nature is neither beautiful nor good and that we do not have to like it. Vojtěch Pelikán will reveal the existence of 'green IT workers'.

This book was researched and written in collaboration with my pupils, Vojtěch Pelikán, Lukáš Kala and Lucie Galčanová Batista. In the chapter on the *Children of the Colourful* (IV.4.), this trio consider – on the basis of field research – whether there is such a thing as an intergenerational transmission of environmental virtues. They also conducted research into the *new Colourful* which poses the question of whether new environmentally friendly ways of life are now taking shape without this generational continuity.[14]

At several points I have inserted poems by the South Bohemian journalist Radek Štěpánek into the text. In a broader sense, Radek is also my pupil[15] because he studied at our Department of Environmental Studies. Are we flattering ourselves too much if we see this scientific grounding as the source for the compelling exactitude of his poems?

The book contains a wealth of 'footnotes'. This is the result of the author's attempt to ensure that the main flow of the text does not leap

[14] The research results are not published in this book; they were written up as journal articles (Kala, Galčanová and Pelikán 2016, 2017).

[15] Radek has explicitly aligned himself with his teacher Zbyněk Ulčák, also a poet, albeit of a different ilk (Ulčák 2015).

from one level of generality to another and to keep it flowing smoothly. My attempt to ensure that the text is logical and easy to navigate – perhaps influenced by a teacher's habits – went even further: it led me to create five *interpretative keys*, located in section II. Here I offer a concise outline of the theoretical concepts that can result in a better and more thorough understanding of the empirical chapters. References to the keys will be indicated by superscript: the universal key,[UK] ontological key,[OK] psychological key,[PK] motivational key[MK] and theological key.[TK] Is it possible to ignore the interpretative keys and skip section II? I have conceived the text in such a way that it is possible, but I think it would be a shame. To make it easier for readers to get to grips with the keys and basic terms, at the behest of my students I am also inserting the CHEAT SHEET prepared by Michaela Kašperová into the book as a bookmark.

The book has one other atypical feature: incorporated into the chapters are thematically relevant TEXTBOXES, often snapshots from life. I hope that the reader won't find them disruptive but will welcome them as something that illustrates and enlivens the material.

At the close of this introduction, it is incumbent on me to return to the subtitle of the book and explain what I mean by the mannerist label of 'ecological foolishness'. Who is foolish and why? The ecopragmatist justifying technical measures by calculating ecosystemic services? Or the poet in despair over a pool that is drying up? Or the activists helping migrating green frogs to cross a motorway? I will leave the answer to the tacit understanding that I hope will develop between the author and the reader in the course of reading the book.

However, with a nod to David Buckel († 14. 4. 2018), I would ask: let us be careful not to use the dismissive, condescending label of 'ecological foolishness' as an excuse stemming from our own comfort, indolence and fear.

II.
Some interpretative keys

II.1. Universal key:[UK] faithfulness, compassion, responsibility

Events in the world prevent us from tearing ourselves away from the news to calmly observe the landscape and the living creatures surrounding us. After all, what's the point? What would we see? Some of 'our places' have disappeared while others are on their way out; it would seem that all we can do is shrug our shoulders. However, some of us have remained faithful to nature and take more of an interest in it precisely because it is weak and facing destruction. Is this faithfulness based on compassion or responsibility? Is environmentally friendly behaviour based on rational considerations or does it arise intuitively? Are rationality[16] and faithfulness contradictory and mutually exclusive? Does it even make sense to think about faithfulness when, in many respects, the relationship between human society and nature has been built on antagonism and on the deceitful abuse of nature?[17] What is the relationship between faithfulness and anthropocentrism?

I am unable to give a satisfactory answer to these questions. Instead they give more of an outline to this book's perspective on environmental issues. I have also offered several interpretative keys to guide the reader. I have termed the first one *universal* as it has either a direct or indirect relationship to the entire book. It has the superscript[UK] which I will subsequently refer to.

The word *faithfulness* is not something we hear in everyday life.[18] It evokes a kind of cautiousness; perhaps because it is filled – it may even seem to us overflowing with – positively charged content. It is used without misgivings in moralising literary genres – religious sermons, fairy tales and legends.[19] I was engaged in a long, futile search through the philosophical and ethical literature until I was rewarded with a mention of faithfulness – to my surprise even in connection with environmental issues – by the philosopher Hans Jonas: "Even if in our devastated environment there was to be a life worthy of the name human for our descendants, the profusion of life on Earth, created by the long-term creative activity of nature, would still be entitled to our protection for its own sake. (...) As those who have been created by nature, we must be faithful to the entirety of its creations" (Jonas 1997, p. 203). Therefore, Hans Jonas's view is reminiscent of the late phase of deep ecology. If we remove the dualistic view from the concept of faithfulness and if we agree with non-anthropocentric deep ecology that we are part of the unity of nature (Næss 1973), then faithfulness

[16] 'Sensibleness', 'rationality' and 'reason' are meant generally as people's mental ability to generalize experience, to apply thought to their decisions and behaviour.

[17] This need not only mean hunting snares, traps and lures. A large part of agricultural produce depends on artifice, for example, the production of milk, crop acceleration, breeding and genetic engineering. The ability of reason to mature in nature is perhaps through perfecting tricks which nature's creatures play on each other. For example, Michael Steele and his colleagues gave a detailed description of how male squirrels deceive their partners when hiding nuts (Steele *et al.* 2008).

[18] The antonym of faithfulness comes up more in conversation and science – deception or betrayal. For example, the perception of truth and deceit in Chinese, Japanese and Jewish traditions was examined by Susan Blum (2007), and "Betrayal and Treachery in the Middle Ages" was a conference theme for historians and linguists in Montpellier in 1995 (Fauré 1997).

[19] Alois Jirásek's *Old Czech Legends* gave us Bruncvík's lion, which as an image of faithfulness became a symbol of the Czech state.

is an essential, unconditional and unproblematic fact. To be faithful to nature simply means being faithful to yourself.

However, the sociological aspect of this book is more complex. Faithfulness is not a state of mind; it is a way of behaving, an action, which was one of the basic chivalric virtues. As often happens as a result of the scepticism within Western culture, faithfulness provoked ridicule. One famous image of absurd faithfulness is *The Ingenious Gentleman Don Quixote of la Mancha*. Elsewhere, in contrast to overly positive and serious heroes, the sympathy of the readers and audiences of comedies is gained by the cunning, dirty tricks of swindlers who quickly adapt to trying circumstances. But it is a strange form of cunning; a cunning when dealing with authorities and the clumsy oafs in power. Despite these and other forms of relativization, there remains a dreadful betrayal of trusting partners, children, lovers and friends, in short, of defenceless creatures.[20] We can already see an ambivalent attitude towards faithfulness in ancient culture. The Greek hero Odysseus dealt with precarious situations using a cunning which we would today term as selfish. The elegant fickleness and unpredictability of the Greek gods, typical of Zeus himself, contrasts with the Covenant of the Lord, which was concluded with the Jewish people. The biblical texts speak of faithfulness as constancy and permanence with a kind of urgency and free of any metaphorical encoding which is otherwise common in the Scriptures. Here faithfulness is not understood as rational or as a benefit derived from loyalty,[21] it is connected to *mercy*.

"I do not conceal your love and your faithfulness from the great assembly" (Psalm 40:11).

"All the paths of the Lord are love and faithfulness, for those who keep his covenant and his testimonies" (Psalm 25:10).

Leaving aside those blinkered scientists who use intellectualization[PK] to trivialize environmental damage, most Europeans agree that nature today is weak and facing destruction. In our culture, chivalric ideals and the Christian emphasis on mercy provide one reason why we should remain faithful to nature. Matt Ferkany (2011) talks about pity as being part of environmental virtues.[MK]

In nature, however, a faithful self-sacrificing feeling for others is not an all-encompassing ascetic universal altruism which derives from burdensome self-denial. In our personal relationships we are able to display self-sacrifice, above all, towards those who we *love*,[22] or to put it more mildly, towards those we are *fond* of – another biblical word. The hidden premise of love is the idea that the object of our faithfulness

[20] The Czech national opera *The Bartered Bride* portrays various forms of faithfulness and deception. Mařenka is thought to have been betrayed, but this is then shown to be untrue. Although the cleverness with which Jeník defeats the marriage broker, Kecal, and the heavy stutterer, Vašek, might be understood as an expression of the questionable morality of Czechs, it is clear that Vašek, who has been deceived, is the preferred son of the rich farmer, Mícha. It is not about a betrayal of trust; it is not about a double-cross: Vašek expects nothing from Jeník. At the end, the chorus celebrates faithfulness. "A good thing has happened; faithful love has conquered."

[21] The term 'loyalty' has historical roots. It was linked to the feudal system, it did not mean faithfulness as a virtue deriving from love, but the result of rational thinking; loyalty was carrying out a serf's duty.

[22] 'Love', like 'faithfulness', has been devalued in a strange way. Due to its emotional character, it is absent from the environmental discourse. However, we come across a 'love of nature' relatively frequently. We shall return to it in Chapter III.1. on the environmental grief of female scientists.

[23] Not necessarily; history also recognizes faithfulness as the blind devotion to cruel rulers.

[24] Erich Fromm points out that it is no coincidence that in the First Book of Moses, God states repeatedly and with increasing intensity that his natural work is good, with an emphasis on the special value of the finished whole (Fromm 1986, pp. 141-142).

[25] We also find this idea in the works of mystics looking at the size of the entire galaxy and the ingenuity of living creatures (*e.g.* Basil of Caesarea, Bernard of Clairvaux, Hildegard of Bingen).

[26] Donald Worster highlights that these optimistic enlighteners found themselves on thin ice when the political philosopher Thomas Hobbes described nature as a place of conflict, fear and violence (p. 45).

[27] In the 1980s, the concept of ecological stability was replaced by an interpretative model based on the principle of competition. However, a significant part of environmental education is still based on a static concept of ecosystems (Plesník and Petříček 2012). It is possible that it is more advantageous for creating a positive attitude towards nature than a dynamic model.

is *good*.[23] We are unable to lovingly care for something which we are afraid of or which disgusts us, such as pathogenic bacteria or parasitical insects.

For aeons people saw nature as something powerful, cruel and evil, dependent as they were upon its whims. We only began to accept nature as a universally shared value when our civilization began to dominate. And yet the idea of a *good nature* is firmly rooted within our culture. According to the Old Testament, "God made the wild animals according to their kinds, the livestock according to their kinds, and all the creatures that move along the ground according to their kinds. "And God saw that it was good" (Genesis 1:25).[24] In 400 AD, St Augustine described ideas in his *Confessions* which can be read today as being ecological in the true sense of the word: "I perceived therefore, and it was manifested to me that Thou madest all things good, (...) and for that Thou madest not all things equal, therefore are all things; because each is good, and altogether very good" (Augustine 2012, p. 156).[25]

In the search for general laws which are valid for nature and society, the thinkers from the Enlightenment admired nature as the good work of God. It appeared to them like a *perfectly run household*. "Nature is such an economist that the most incongruous animals can avail themselves of each other!" wrote the pastor and naturalist Gilbert White in the 18th century (cited from Worster 1985, p. 8). In "nature's economy" they valued the precise mechanical functionality of the Creator's works and tried to connect it to the idea of God's kindness to the beings He created (Worster 1985, p. 44).[26]

This line of thought reached its pinnacle in the 20th century with the ecosystem ecology formulated by E.P. Odum (1953). He stated that nature represents a system which, through a cycle of substances and energy, develops into a permanent, species-rich state. In classic ecosystem ecology, ecosystem stability is implicitly understood as desirable and good.[27]

If we ignore the views of scientists, our intuitive understanding of 'good nature' is influenced by another line of thought: The connection between nature and goodness is aesthetic and spiritual in character, either ignoring or forgiving nature for its so-called 'natural evil' (Chapter III.4.) based on predation and competition.

From the period of Late Antiquity to the present day, the love of good nature has been derived from admiring its *beauty* (Stibral 2005). Idealistic philosophy, influenced by the ideas of J.G. Herder and J.W. Goethe, differed from Enlightened mechanical reductionism and causal interpretation in that it emphasized the depth and value of natural existence. Nature became a romantic expression of the aristocracy's interests and an emblem of national self-confidence. When within this field of thought, reinforced by the process of individualization (Librová 2010), industry expanded into the landscape and drastically destroyed the beauty of nature, the result was clear: The protection of nature was born as a cultural phenomenon in which *compassion* played an ever greater role.

Despite its deep cultural roots, the principle of compassion towards nature is not taken for granted, even within environmental ethics. The egalitarian stream of criticism highlights the *anthropocentric* character of compassion. Compassion assumes a relationship of dominance and *benevolence* (Frasz 2005), and does not, therefore, concur with the convictions of deep ecology on the unity of all beings, nor with the outlook of some branches of ecofeminism which view compassion as an expression of the patriarchal conception of the world (Librová 1994, pp. 81–88; Šmídová 2009). Other egalitarian doubts about the adequacy of compassion with suffering nature lie in – empirically based – evidence on the blurring of the boundaries between people and animals, which has recently been examined in a series of scientific works both from the natural sciences (Lehečková 2016) and theology (Deane-Drummond 2014).[28]

For the purposes of this book, important doubts concerning compassion have been raised by the political scientist Andrew Dobson. He disregards the idea that compassion is a suitable response to greatness of today's environmental problems. He proposes replacing the charitable emotional concept of 'the Good Samaritan', guided by ethical virtues and duties,[MK] with the rational principle of *environmental responsibility* and the principle of the 'environmental citizen'.[29] Such a citizen makes free rational decisions as a guilty party, aware that their ecological footprint is so great that it is beyond the planet's endurance. Whoever benefits from the impoverishment of the Earth should turn to more responsible behaviour to reduce their consumption (Jonas 1997; Dobson 2003, p. 54). However, is it possible to find empirical evidence of the environmental responsibility of this responsible citizen in reality?

The first step in the search for an answer might be the differentiation between *the ethic of ultimate ends* and *the ethic of responsibility* as proposed by Max Weber (1946, pp. 41–48). We often come across the ethic of ultimate ends in the rhetorical field. It is usually an aspiration

[28] In the theologically orientated Chapter II.5., I also express doubts about compassion being the basis of the relationship between humans and nature. We look on with a pastoral compassion as nature suffers, while it is still possible to hear the intuitive folk wisdom of 'nature will help itself'. I believe that the sibling relationship, or a relationship of responsibility, is more appropriate – principles which are supposed to prevent ecological damage.

[29] The theme of environmental responsibility has been examined extensively in the academic literature. In November 2009, Lukáš Kala found 241 articles containing the phrase 'environmental responsibility' in the Elsevier Scopus database. However, he doubted the epistemological value of these texts (Kala 2009, p. 16).

which is more an expression of wishful thinking. The ethic of responsibility is more matter-of-fact: it rationally accepts that people are not perfect. In the case of environmental problems, the difference between the ethic of ultimate ends and the ethic of responsibility is particularly important. It is evident in the conflict between the generally shared value of nature, the genuine attitude of 'I love nature', and everyday environmentally damaging behaviour.

It is debatable whether an individual, no matter how responsible, can reduce today's advanced state of ecological damage or transform a bleak ecological future. Michael Maniates has doubted the effect of responsibility if it is understood as individualized lifestyle, in particular 'green consumption'. In his article with the ironic title *Individualization: Plant a Tree, Buy a Bike, Save the World?* (2001), he criticizes the deceptive shift of environmental responsibility from society's social and economic structures to the individual.

Faithfulness to the nest

At the end of April, the swifts return after a long flight from central and southern Africa. From dawn to dusk, all summer long, the atmosphere in Brno is filled with the cries of these daredevil fliers, skimming over the rooftops. They seem to be calling out 'see-zee, see-zee'.

The couples don't fly from their winter resting place together but meet at the nesting site. Scientists are still baffled by the fact that after such a long journey they are able to unerringly find their own city, house and nest. These are usually built where the brick is crumbling from the ledge of an old house, in a cavity where a piece of plaster has fallen from below the eaves, or behind the ventilation space in the roof area of a block of flats. The faithfulness of swifts to their nest is legendary – as impressive as it is touching. Ringing the birds has shown that swifts will return to their nest over a period of up to eighteen years – the whole of their relatively long life. Swifts do not like having to find new nesting sites.

This is a disadvantage for conservatives in a dynamic world. The homes in our country have changed radically over the past twenty years. Crumbling facades and dilapidated roofs have quickly disappeared. Attic conversions have become more common and blocks of flats have been 'ecologically' insulated. The result is an ecological paradox: the nesting opportunities for swifts have declined dramatically.

In May this year, I watched as some swifts searched in vain for their nests in a renovated courtyard, and how they finally resigned themselves to finding an alternative place under the two remaining old roofs. In June, I saw the renovation work of destruction completed: scaffolding was placed around both houses and the roofers began removing the tiling. I don't want to think about whether they threw out the nests with the eggs or with the chicks already hatched.

Swifts are strictly protected and you are not allowed to carry out any activity which could endanger them while they are nesting. But who would take the law seriously in this case? Construction workers have to earn money and the owners want to keep their property in order. Some owners, though, would happily leave the swifts under the roofs and live alongside them. The Czech Society for Ornithology, which has been observing swifts in towns over a long period, is able to advise these people. The first condition is to make sure in advance that construction work does not interfere with the nesting. Then there should be discussions with the project manager on how to provide new opportunities for the swifts to nest in the following years, for example, using specially constructed boxes. Whether such conservative birds will actually nest there is another matter.

II.2. Ontological key:[OK] vita activa and vita contemplativa

[30] These are meant as types: it is impossible to include every human activity within them, let alone all professions.

[31] This concept was not introduced into philosophy by Hannah Arendt herself, even though her name is often linked to it. The concept 'vita activa' appears in 1,005 entries in a database of medieval Latin Christian authors in connection with 'vita contemplativa', which will be discussed later.

[32] Arendt described a caring person as *animal laborans*, which indicates a similarity to the activities of animals who do not create anything new. After consultation with Latin and medieval scholars, I propose replacing the adjective *laborans* with the word *curans*.

[33] In *The Colourful and the Green* I wrote about how the physicist Christian Schütze saw the destructive effects of human manufacturing activities on nature. When in the subtitle of his book, which is based on the second law of thermodynamics, he asks "Is work ruining the world?" (Schütze 1989), he is thinking about *work*. Leaving aside the different terminology and different approaches of the disciplines, the statements by the two authors are analogous.

[34] This category of activity is closely related to the ethical motivation which is described as 'teleology' in Chapter II.4.

[35] The fact that this idea is only presumed is an important fact regarding environmental issues.

In order to explore environmental issues in more depth, it might be useful to take into consideration Hannah Arendt's analysis of human behaviour. We will keep to categories which are based on her analysis,[30] while maintaining their Latin wording, as it keeps all the social, psychological and personal connotations due to its abstract form. The set Latin terminology reduces the danger of misunderstandings which are common in laymen's philosophical discussions – 'one is talking about a coat, the other about a goat'. There is a dictionary of the Latin terms in the second TEXTBOX of this chapter.

In her extensive study *The Human Condition* (1998), which was first published in 1958, Hannah Arendt saw the active life[31] as a combined term for three basic types of human activities, which she describes as *labour, work* and *action.*

These activities are mainly cyclical in character, satisfying people's needs through goods which are quickly consumed or they disappear. For this book it is important that 'labour' contains important aspects of *faithfulness.* Examples include household work, looking after a child or a sick person, and, of course, protecting the environment. A person who maintains something through their work can be described as *homo curans* (caring person).[32] Arendt emphasizes the fact that in the Bible the description of Adam was of someone who laboured in the fields. However, "the *animal laborans*, which with its body and the help of tame animals nourishes life, may be the lord and master of all living creatures, but he still remains the servant of nature and the earth" (1998, p. 139).

According to Arendt, only *homo faber* – a toolmaker that transforms natural matter into a useful product – conducts himself as lord and master of the whole Earth. "Human productivity was by definition bound to result in a Promethean revolt because it could erect a man-made world only after destroying part of God-created nature" (p. 139).[33] *Work* has a clear purpose and goal,[34] which is known in advance and the attainment of which is in the hands of an expert; he has, we believe,[35] the necessary means and methods to reach the desired goal. Henri Bergson understood the intellect as "the faculty of manufacturing artificial objects, especially tools to make tools" (1919, p. 193). The manufacture of tools and machines and the act of experimentation are at the very essence of the modern-age revolution. The dismissive modern-age view of labour was expressed succinctly by

Adam Smith: He understood domestic labour as a form of parasitism because it does not increase the world's supply of things. Smith was scornful of the "unproductive nobility", and even of the "menial servants", who like "idle guests (...) leave nothing behind them in return for their consumption" (Smith 1976, p. 338).

We might recall the proud, confident expressions on the faces of developers and architects as they stand beside investors and respected journalists during a site survey of a wetland meadow or blossoming steppe vegetation. The objections raised by the obligatory conservationists are pitifully weak against the arguments of the engineers. The project leaders – *homines fabri* – planning a motorway propose a piece of "WORK"; at the moment there is "NOTHING".

Another illustration is offered by the American science-fiction film *Interstellar* (Nolan 2014). It opens with scenes of withered fields of corn infested with resistant mould. Above the land hovers an omnipresent cloud of dry dust, threatening to bring famine. Time is running out for humans on the planet Earth. In order to avert a catastrophe, a group of researchers aim to colonize a planet in a distant galaxy. The hero of the film, a corn farmer, is now a pilot/astronaut faced with a difficult decision: Should he try to help save the dying Earth or should he set out on a radically new technological adventure? He opts for the intergalactic journey and complains when some people procrastinate: "It's like we've forgotten who we are, Donald – explorers, pioneers, not caretakers."[36]

And one more example of how nature conservation is perceived: A citizen shakes his head as he watches an environmental protest on the news. The young men and women attached to trees or blocking a road elicit the parental admonition: "Why can't they do something proper?!" What is this 'proper' thing in the mind of this respectable citizen that young people should be doing? They should be baking bread, assembling car parts or designing aeroplanes, introducing any kind of product to the world which is truly or supposedly useful. Since the start of the Modern Age to the present day, this is how the respectable and commendable activities carried out by *homo faber* have been perceived.[37] Nature conservation does not belong to this 'proper' category.[38]

The third dimension of active life is *action* (*homo agens*). By this Arendt means the actions and discourse of a human as a political being (the Aristotelian *zoon politikon*). It is not due to necessity, as is the case for labour; neither is it linked to the usefulness which is the

[36] In the US the word 'caretaker' is used specially for nature conservationists.

[37] However, this idea does not reflect reality today – workers in the service industry outnumber producers (ČSÚ 2015). A typical job for the first decades of the 21st century would be an IT specialist. In the 1950s, Arendt bitterly stated that the society of work and labour was transforming into a society of employees, with the result that people were losing experience of the outside world.

[38] But let us not oversimplify. Part of the public would see the activists as useful if they cleaned wells, planted trees and did not cause problems: "I've nothing against nature or environmentalists – as long as they are *reasonable.*"

result of work. Action is often branded as 'idle uselessness' (Arendt 1998, p. 220). It does not bring material results, but it is no less real than work or labour. However erroneous it may be to apply the measurement of *homo faber* to action, it is also necessary to acknowledge that action, "this power, which is achieved by a large number of freely acting individuals, who seize the initiative and are capable of organizing themselves in such a way that they increase the action potential of each person, is irreplaceable, and it is precisely this which enables the group to acquire real influence and a share in the decision-making on public affairs and 'important matters' concerning its own polis or country" (Němec 2010).

In Ancient Greece neither labour nor work were recognized as fulfilling the life of a free citizen. A clear distinction was made between the domestic sphere and the public area of the *polis*, which was the sphere of action. However, the true ideal of ancient wisdom was something else: *a contemplative life – vita contemplativa*: a noble pastime, the creative activity of understanding the truth and essence of being. The philosopher – *homo contemplans* – was freed from the burden of public life.

Many thoughtful environmentalists, followers of deep ecology and countryside ramblers view the mode of contemplation as being the opposite of an active life. However, Arendt should not be confused with the position of radical observers and lovers of nature, inspired by the meditation techniques of Eastern cultures. She is writing about changes to the meaning of the contemplative mode in European culture, and how until the start of the Modern Age *vita activa* had been viewed with suspicion and contained the connotation of 'un-quiet' (Arendt 1998, p. 15). Thomas Aquinas also believed that peace could not be found in an active life. The meaning of life could not be through acquiring material things; the contemplative life had priority – an idea which became popular through the New Testament story of Luke's Gospel (10:38–42) about the visit of Jesus Christ to the sisters Martha and Mary.

According to Aquinas, a contemplative life is filled with joy because its subject is the presence of God and the contemplation of his reflection in creation, which points to inexhaustible depth and beauty. At the conclusion of her book, Arendt writes in a similar spirit and with a critical stance towards modern-age views: *homo faber's* "greatest desire, the desire for permanence and immortality, cannot be fulfilled by his doings, but only when he realizes that the beautiful and eternal cannot be made" (p. 303).

However, from an environmental perspective it is significant that Aquinas did not assert the primacy of the contemplative life. In the New Testament he encounters the fundamental importance of the active love of your neighbour. The *vita mixta – mixed life* began to develop in monasteries, in particular with the Benedictines (Scherer 2005). In *The Colourful and the Green* I wrote from an environmental perspective on specific human qualities and need to transcend his being (Librová 1994, pp. 21–23). It is a position which goes beyond the framework of everyday reproduction (*i.e.* labour/care), a self-extension that responds to a deficit in life, to the awareness of mortality. The category of transgression, which is of key importance for our reflections, correlates approximately with Hannah Arendt's analytical categories: "Transgression in things" corresponds to the *homo faber* mode of work. "Transgression in people" corresponds to the mode of action, but also, in the care of others, to the mode of *animal curans*. "Transgression in symbols" can be classified together with the contemplative life (Petrusek 1986).

What role do these three modes of active behaviour play in environmental protection and in the everyday reality of an environmentally friendly lifestyle? What role does contemplation have to play within them? These are questions we will touch on in part III.

Mary has chosen what is better

As Jesus and his disciples were on their way, he came to a village where a woman named Martha opened her home to him. She had a sister called Mary, who sat at the Lord's feet listening to what he said. But Martha was distracted by all the preparations that had to be made. She came to him and asked, "Lord, don't you care that my sister has left me to do the work by myself? Tell her to help me!"

"Martha, Martha," the Lord answered, "you are worried and upset about many things, but few things are needed—or indeed only one. Mary has chosen what is better, and it will not be taken away from her."

(Luke 10:38-42)

Vocabulary based on Hannah Arendt's terms

homo faber (man the maker), pl. *homines fabri*
animal curans (caring animal), pl. *animalia curantia*
homo curans (caring man), pl. *homines curantes*
homo agens (active man), pl. *homines agentes*
homo contemplans (contemplative man),
pl. *homines contemplantes*
homo contemplator (man capable of contemplation),
pl. *homines contemplatores*

II.3. Psychological key:[PK] defence mechanisms, locus of control

[39] These are also referred to in the literature as 'coping strategies'.

[40] Psychologists advise environmentalists not to scare people when communicating with the public and to avoid negative information; studies have shown that it often leads to feelings of guilt and fear and reinforces defences rather than environmentally friendly attitudes (Krajhanzl 2014, p. 131). However, this psychological fact is at odds with the philosophical view of Hans Jonas, which is reproduced in the introduction.

[41] For example, in Australia and in Scandinavia.

In the introduction I briefly looked at one of the obstacles standing in the way of environmental awareness. I called it *evolutionary deafness*. Surprisingly, those who have overcome it do not experience a feeling of relief from having seen the light; on the contrary, the consciousness of environmental threats becomes a burden that leads to a troublesome mental state. They come to realize the magnitude of environmental problems as well as the difficulty in solving them. Because individuals sense that they are partly to blame, they consciously or unconsciously feel reports about the environmental crisis to be an attack on their psychological well-being and inner integrity – the integrity of the ego. In order to escape this unpleasant state of mind, they resort to various defensive strategies.[39]

Psychologists and psychiatrists are able to help people afflicted by personal misfortune, but what advice can they give to people suffering from a fear of something as difficult to grasp and difficult to solve as an ecological crisis?[40] This question may seem exaggerated or premature to some readers, but in recent years defences and strategies for dealing with environmental problems have been the topic of discussions in the fields of ecopsychology and cognitive and analytical psychology as well as the subject of psychological counselling in countries badly affected by climate change.[41]

The psychologists Hamilton and Kasser (2009) differentiate between *maladaptive coping strategies*, which question or in some way undermine the importance of negative information, and *adaptive coping strategies*, which accept the information and lead to a change in behaviour, to an attempt to put things right. As we will see in the empirical part of this book, between maladaptive and adaptive coping strategies there is also a wide range of crossovers, combinations and difficult-to-categorize cases.

Adaptive and maladaptive defences

At the heart of this book is an environmentally friendly way of life. In the language of psychology, taken with a pinch of salt, what we will be concerned with is a life founded on adaptive defences. The reader will be able to join the researchers in observing how various forms of *altruism* are characteristic of the Colourful: for example, working for the community. Another form of adaptive defence is *sublimation*, in which repressed emotions such as anxiety about developments in the world are transformed into works of art.

However, in this complex world full of negative information and difficult situations, it isn't possible to get by in the role of positive heroes. The Colourful employ some maladaptive defences to help them in their everyday lives. In one way or another, they use them – for the most part unconsciously – on a daily basis. It is the very diversity of maladaptive coping strategies used that makes the Colourful group so interesting, vital and compelling.

It is not easy for an analyst untrained in psychological systematics to get to grips with the structure of maladaptive defences. I eventually opted for the classification proposed by Clive Hamilton and Tim Kasser (2009). I have supplemented their view, discussed at a climate change conference in Oxford, with the concept of defences outlined by Deborah Du Nann Winter and Susan M. Koger (2010).

People can respond to negative environmental information by *denial (1)*, a mechanism that allows us to insist that the anxiety-provoking phenomena simply do not exist. This defence mechanism can be ostensibly backed by science, as so-called climate sceptics and deniers have attempted to do. In the Czech context, a legendary figure among them is the former president Václav Klaus: "Global warming is a myth and I think that every serious person and scientist says so" (Roberts 2007).[42] Similar defence mechanisms include the unconscious process of *repression*, moving information from the conscious part of the mind to the unconscious, as well as conscious *suppression*, which aims to shift attention to something else.

Consistent forms of defences are actually quite rare. A more common maladaptive defence mechanism in relation to disturbing information is *reinterpreting the threat (2)*: The individual acknowledges environmental problems but underestimates their seriousness; they fragment reality, only paying attention to the part of it that can be construed in a relatively favourable way.

This form of defence is very familiar from the environmental sphere, for example in the comforting reminder that "scientists are often wrong". Reinterpreting the threat also makes use of the opposite argument: People brought up in the Enlightenment spirit of faith in science are happy to accept as an argument the simplistic assertion that "Humans have solved these sorts of problems before" and reassurance from experts about the omnipotence of smart technological solutions.

[42] A book summing up Klaus's climate-sceptic line of reasoning was published in English under the name *Blue Planet in Green Shackles* (Klaus 2008).

Some experts in the natural sciences support reinterpreting the threat through another type of defence known as *intellectualization (3)*. They deliberately distance themselves from the sensory, emotional and value components of knowledge and try to create emotional distance by means of abstraction. Intellectualization makes use of numbers, measurements and statistics, often without compelling justification and based on kneejerk reactions. This type of defence is illustrated by the arguments of Bjørn Lomborg, which we are familiar with from his book *The Skeptical Environmentalist* (2001).

Distancing (4) can be considered a form of maladaptive defence where we acknowledge the seriousness of the environmental situation but defer its manifestations, repercussions and solutions until the future – the familiar refrain of "it won't affect us" or even, with a touch of cynicism, "that's something our grandchildren will have to deal with".

Hamilton and Kasser draw attention to one relatively widespread type of maladaptive defence which they refer to as *diversionary strategies (5)*. It consists of activities that attempt to alleviate the feeling of guilt without requiring a fundamental lifestyle change. These are familiar measures included in lists of 'tips to help the planet': "Turn off the tap when shaving", "lower the thermostat in your house" and so on.

While these small-scale changes in everyday behaviour do, of course, produce some kind of beneficial effect, whether real or symbolic, there is a type of maladaptive lifestyle defence which is harmful to the environment. The individual aims to liberate themself from the feeling of threat and anxiety through *pleasure-seeking (6)* in the form of shopping, exciting experiences, adrenaline sports or activities that can be termed consumerist.

One interesting maladaptive defensive manoeuvre on the part of our psyche is *reaction formation (7)*. In general terms, a person defending their ego through reaction formation is convinced that somebody wants to harm them and resorts to a feeling of hostility. Reaction formation is part of biophobia (Orr 1994), a fear of nature. In an age of natural disasters, we hear quips with an ambivalent subtext like "nature is paying us back" or "nature is taking revenge on us". On a more general philosophical level, there is the emergence of the theme of nature's evil side, its cruelty and the need to tame it or refine it.

One specific example of a maladaptive defence is *aggression (8)*. This is conspicuous in the attitude of a section of the Czech public to environmental activists who disturb their sense of everyday security with worrying information, reminders of ecological guilt and moralizing. The public find protests by environmentalists particularly irritating. By directly or indirectly calling for a change in everyday behaviour, they apparently threaten the individual's freedom to decide how to live.

In empirical research we have repeatedly come across a type of defence which is not mentioned in the literature. When the subject of the critical state of nature and the poor prospects for the planet's resurgence came up in interviews, some of the respondents expressly attributed their peace of mind to the simple fact that they are religious. They agreed with the questioner's formulation that they were resorting to *defence by faith (9)*. It is not based on the conviction that God is going to solve the current ecological crisis for us, but on counting on the development, whose outcome we cannot yet glimpse. It is a defence grounded in a sense of complete trust and reliance. As our research has shown, this type of maladaptive defence often evolves into an adaptive defence, into active pro-environmental behaviour.

There is another type of maladaptive defence which can develop in a similar way over time, *i.e.* towards action – *affectualization (10)*. It is an attempt to defend the ego through an emotionally charged psychological state, through environmental grief,[43] by means of regret and sorrow.

A common type of defence in the environmental sphere is *rationalization (11)*, [KU 44] attempting to explain something that is hard to concede in an acceptable way; for example, by referring to a 'losing battle'. This is a response to major threats such as global climate change which the individual cannot influence.[45] There is also a rationalizing basis to attempts to assign blame to others: "Others have a bigger ecological footprint than I do," "China's building one coal-fired power plant after another," "The Czech Republic's too small to make a difference." At first sight rationalization comes across as an adaptive attitude. It seems rational to give up on attempts to rectify environmental damage. And yet it is actually a maladaptive defence because it distorts and diminishes reality by neglecting the value, ethical and emotional aspects.

Locus of control

Whatever endeavour we might embark upon, it is important whether or not we feel strong enough to see it through. At the same time, we ask ourselves the question: Will we come up against external factors that will thwart our plans? In psychology these questions are dealt with by the concept of *locus of control* (Rotter 1954). According to Julian B. Rotter, people with an *external locus of control* take the view that they themselves cannot influence things, that external factors matter more than their own efforts. By contrast, people with an *internal locus of control* are convinced that they are decisive agents of change and are able to influence the world around them. In addition to a person's innate psychological make-up, an individual's life experience and the information they have received can also play a vital role in shaping their perception of the locus of control. In Chapter III.1. we will become acquainted with Paul Kingsnorth and his associates, who after many years of disappointment at the failures of environmental activism became convinced of the insurmountable predominance of the external locus.

[43] Because environmental grief can play a part in shaping environmentally friendly behaviour and because it is a direct expression of a faithful attitude towards nature, a separate chapter will be devoted to it (III.1.).

[44] We should not confuse this psychological aspect with the principle of rationality as it is conceived in the history of philosophy. There rationality is regarded as one of the main causes of the environmental crisis. The most frequently cited work is a book by members of the Frankfurt school, Max Horkheimer and Theodor W. Adorno, *Dialectic of Enlightenment* (1944). It sees the source of man's alienation from nature as 'modern rationality' and the 17th's and 18th's centuries 'spirit of Enlightenment'. However, its roots go back to Olympic mythology, to the story of Prometheus.

[45] See the external locus of control.

The issue of the locus of control is largely a topic from health psychology and health education. Studies have confirmed that a belief in an internal locus of control can be an important precondition for recovery and part of successful prevention (Rodin and Langer 1977). When it comes to environmental problems, matters are more complicated. Unless we happen to be ecopragmatists,[46] we will have doubts about the power of the internal locus.

In the environmental sphere, an internal locus of control can benefit implementation when it comes to clear and simple environmental issues at the local level (Pavalache-Ilie and Unianu 2012). If an individual or a group of people decides to clean up a natural spring or plant an alley without outside help, their plan is likely to succeed; the attitude of an internal locus of control corresponds to reality and has a positive effect on the implementation of the project. However, when the same person decides that the council should clean up the local pond, the view of an internal locus will no longer be so unequivocally valid. Success will also depend on a number of external factors of a sociological, administrative, legal, financial and technical nature. Ultimately, when it comes to solving complex global environmental problems, an external locus of control is appropriate. The belief in an internal locus of control is by its nature an ethic of ultimate ends.[UK] An attempt by a single household to reduce car journeys cannot influence changes to the planetary climate.

[46] See Chapter III.2.

Seminar students on the faculty terrace.

Seminar Study A

Students attending my 'Lifestyle and Environmental Issues' seminar carried out two small-scale research surveys in the spring semesters of 2014/2015 and 2015/2016. Survey A focused on psychological categories of defences. The students assembled a set of at least ten people whose characteristics roughly corresponded to the structure of the Czech population and used it to verify the empirical appropriateness of the theoretical types of defences in relation to disturbing environmental information. They were to estimate the frequency of the defences.

The defence mechanisms most frequently given by the respondents were *diversionary strategies* ("I recycle and save water", "I don't buy palm oil") and *rationalization* ("I'm not to blame – it's the fault of those in charge of companies, the government and regional officials; there's nothing I can do about it"). Respondents often defended themselves using *distancing in time*. One female respondent (a university-educated biologist) reacted *affectionally* – by crying.

The students' inquiries, presented in seminar papers in spring 2015, contributed to the preparations for the interviews with the Colourful twenty-three years on. They allowed us to refine and better operationalize the research questions, confirmed the viability and applicability of the theoretical categories of defences and indicated their incompleteness (the psychology literature does not mention the category of *defence by faith*).

II.4. Motivational key:[MK] teleology, deontology, virtue ethics

[47] Instead of 'purposive rationality', the term usually given as an approximate synonym is 'instrumental rationality', which puts the emphasis on efficiency in achieving the goal.

[48] Purposive rationality, which is the foundation of modern technology, is symptomatically attacked from environmentalist positions as the source of the ecological crisis.

What is the source of a faithful attitude towards nature? Why is it that some people refuse to give up the struggle, although it would be easy and – given the way things are going – reasonable to do so? Why does someone trudge through wet snow to check on the status of black grouse? Why does someone opt for vegetarianism even though they like the taste of meat? And why do others give in? How do two such different attitudes arise?

There are a number of general sociological and psychological theories that can be used as a basis for reflecting on the environmental motivations for faithfulness and reasonableness. I have selected two theoretical sources which complement one another.

First let us briefly turn our attention to the well-known typology of action established by the sociologist Max Weber (1922b). Real behaviour is a combination of four ideal types. *Rational-purposeful action*[47] is directed solely towards a goal. It makes use of objects and people as means for its own rational purposes with a view to success. This type of behaviour considers means in relation to purposes, purposes in relation to side effects and various possible purposes in relation to each other (Skovajsa 2012, p. 581).[48] *Value-rational action* is based on certain values which are regarded by the actor as more important than achieving any goal. The actor remains faithful to the value, however unsuccessful they might be in attaining it. They try to achieve the goal using rational means, but from the perspective of purposive rationality this type of behaviour is regarded as irrational (Pechová 2001). *Affectional action* is based on the emotional states of the actor; it represents impulsive, non-rational responses to stimuli, and – as with value-rational action – its significance does not lie in its success. *Traditional action* does not have a rational basis either. It is founded on habits and the fact that this is simply the way it is done or has always been done.

The Weberian typology of action outlined here loosely correlates with the typology of motivations which will form the main framework of our own categorization. It was derived for the purposes of environmental theory on the basis of normative ethics by the economist John O'Neill and the philosophers Alan Holland and Andrew Light (O'Neill *et al.* 2008, p. 82). They distinguish between *teleological* and *deontological* motivation and motivation consisting of *virtue ethics*. It goes without saying that in practice these three types merge and combine with each other.

Teleology (*telos* = goal) considers the measure of an action's appropriateness to be the ends pursued. The advantage of the teleological approach is its factuality and rationality, the connection with concrete reality; on the basis of observation and experience, we can react to developments, continually monitor the appropriateness of our decisions and correct them where necessary. Teleological motivations for behaviour focus on the future and are not bound by norms grounded in tradition. They are typical of a modern society whose institutions and individuals have adopted the viewpoint of independent agents freely making decisions about the goals of their behaviour. Teleologically motivated behaviour is founded on a belief in an internal locus of control.[PK]

In the environmental sphere, the teleological approach – like the purposive rationality formulated by Weber – is characteristic of technological, political and legal solutions implemented or backed by institutions. However, teleological motivation is also present in the environmentally friendly behaviour of individuals: "I'm going to insulate my façade and get solar panels *to* reduce the ecological footprint of my household," "We have to live more modestly *to* prevent disaster, *to* save the planet," "I'm going to take part in a blockade in the Šumava mountains *to* preserve endangered ecosystems," "We're campaigning *to* save an alley of linden trees." Note that the sociological observation of motivations can take advantage of the fact that lived teleology often uses the preposition 'to' (in the sense of the phrase 'in order to do something').

Anthropologists and psychologists point out that people only *appear* to be reasonable and rational in setting goals. They are limited by non-rational psychological predispositions (*e.g.* Wuketits 1998). The tendency to think in simple terms of cause-and-effect underestimates unintended side effects, which are often unexpected and yet extremely important, particularly in the sphere of environmental problems.

Teleology – which, as the child of utilitarianism, underpinned the achievements of Western civilization – evokes an optimistic notion of simple solutions. Thanks to its purposefulness and scientific foundation, it unquestionably produces small-scale successes; however, it is often unable to fully appreciate the nature and extent of problems. It underestimates external factors of an economic, political and social nature.

And that's not all: teleologically based efforts primarily run up against the character of natural processes themselves – their unpredictability, full of feedback mechanisms. People formulate their civilizational goals in the short term, while natural processes mostly develop at a slow pace (Orr 1996) and often surprise us with a sudden leap. So nature does not always respond positively to the helping hand deliberately extended by man. It is difficult, if not impossible, to estimate when this might happen.[49] The disconnection between simply and partially defined goals and the complexity of natural processes cannot even be eliminated by systematically incorporating scientific knowledge and inviting participation from experts. The fact that modern science is not able to give straightforward advice on human behaviour in complex environmental issues is illustrated by the case of the Šumava National Park, which will be mentioned in more detail in Chapter III.3. There is also a reminder of it in the TEXTBOX for Chapter III.2. "An attempt at creating steppe land".

Moreover, it is worth noting that in the environmental sphere, supposedly rational teleological decision-making is often directly or indirectly guided by fear of the way things are going.[50] This outlook further constricts decision-making about goals in terms of scope and time scale (Lorenz 1974, pp. 27–28), thus depriving teleologically motivated behaviour of its strength, its rationality and detached view.[51] Concurrently and paradoxically, the teleologically motivated environmental activist is the descendant of those who built the successful Western civilization. They believe that the threats they fear can be averted by achieving the goals they have set. They have great expectations.

We should not be surprised that at times of broader and deeper reflection, the teleologically oriented environmentalist feels overwhelmed by the vast scale of ecological problems. They see the success of their efforts as insignificant compared to the overall course of the planet. They say to themselves: "While I'm plugging away in a minor conservation cause, humanity's ecological footprint is rapidly growing, climate change is intensifying and biodiversity is declining." A burnt-out environmentalist often abandons their previous standpoint of an internal locus of control and in resignation turns to an external locus.[PK]

The opposite of teleology is often considered to be **deontology** (*deon estin* = it is necessary). While teleological decisions are made by an independent agent, deontologically based behaviour is founded on tradition. It corresponds to Weberian 'traditional action'. It is based on duty and commitment, on social norms that determine how we should or should not act. The advantage as well as the disadvantage

---

[49] Field ecologists could cite dozens of examples from their own experience. It is worth mentioning two well-known cases: a small carnivorous mammal called the Javan mongoose (*Herpestes javanicus*), originally from Iran, was introduced to many islands in the Pacific Ocean with the aim of eliminating non-native rats. However, in time it began to decimate the local bird and reptile populations (Hays and Conant 2007). Another well-known example is the rose *Rosa multiflora*, which was planted in the eastern USA as a form of erosion control and a refuge for wildlife. However, this plant also began to behave invasively and crowd out native species (Amrine 2002).

[50] Examples of this approach include the famous calls to action from Al Gore (2006) and Bill McKibben (2011).

[51] However, compare this with the view of Hans Jonas, who considers fear to be a condition for environmental responsibility (Chapter II.1.).

of deontologically founded attitudes is their long duration and low variability – in a word, *faithfulness*. If the conjunction *to* is often used to express a teleological motive, in the case of a deontological attitude we encounter the conjunction *because*.[52]

For thousands of years, the norms of the European relationship to nature were derived from the need to push nature back. It was not until the technological power of humans increased that this attitude proved to be detrimental. An awareness of the value of nature took shape, the attitude that we call a love of nature came into being, and ideas and institutions to protect nature were established. A recognition of the beauty of nature, especially the beauty of the landscape, played a crucial role in this. Initially, a non-economic interest in nature was a pastime of the nobility and a subject for works of art. Gradually, through the influence of Romanticism, it became a fashion and a social norm in some circles: We must protect nature *because* it is beautiful, or *because* it is good (Stibral 2005).[UK] Fairly recently, in response to the expansion of technical and urban civilization and to the visible damage done to nature, wilderness – unspoiled nature – has become a rare commodity and has been ascribed intrinsic value. Hence the modern attitude: wilderness should be protected. However, despite its historical roots, respect for nature has still not acquired the character of a strong and binding norm. In particular, the value of wilderness remains socially specific.[53]

A characteristic aspect of deontological motivation is that it appeals to conservatively minded environmentalists. Roger Scruton refers to the existence of an order, the 'hereditary principle' of Edmund Burke, and finds a common basis for the motivations of conservatives and environmentalists: we must protect nature *because* it is our home and the landscape of our ancestors (Scruton 2006). Some environmentally relevant social norms are the by-product of socially inherited practices. Traditional values, founded on the experience of previous generations, are based, for example, on the ancient deontological element of a way of life enjoining thriftiness; there is a particular emphasis on not wasting food.

Conspicuously deontological motivation is contained in the Christian attitude to the world. For example, according to Thomas Richard Dunlap (2006), the attempt to protect nature and live a 'green' lifestyle is grounded – often without thinking about it – in a belief in the existence of an invisible order; our highest good consists of aligning ourselves harmoniously with this order.

[52] Similarly, the verbs 'should', 'be obliged to' and 'have to' are typically used to express a deontological attitude.

[53] With regard to deontological motivations in the environmental dimension, the following should also be noted. Although ethicists relate deontological motivations to the experience of previous generations, in modern times the expert opinion of scientific specialists, mostly conveyed in a simplified and tendentious way by the media, plays a major role in creating norms.

An important religiously based, albeit unconscious, deontological element in the motivation for nature conservation is solidarity with the disadvantaged. The concept of nature conservation as interspecies altruism can be seen as an extension of the duty of compassion,[TK] which is the foundation of a culture shaped by Christian values over thousands of years. I must protect nature *because* it is weak and oppressed.[54]

When modern theology directly addresses the relationship between humankind and nature, it emphasizes that nature is a gift and not something to be taken for granted, evoking awe at its greatness and perfection. The result is a deontological attitude, a commitment, an obligation on the part of the believer to feel gratitude for creation. I treat nature with care *because* I must be grateful to God – its Creator – for it. Miranda Harris, co-founder of the international Christian conservation organization A Rocha, said in an interview for the environmental magazine *Sedmá generace* (The Seventh Generation, Balharová 2012, p. 13): "I'm not sure Christians can hope to reverse the terrible damage being inflicted on Earth. But we try to care for creation because it's the right thing to do, not because it will save the planet." She thus distanced herself from teleological motivations and inclined towards deontological motivations. When she went on to describe the conservation work done by Christians as "an expression of obedience", she was echoing Roger Scruton, who sees obedience as a condition for creating freedom, a basic characteristic of humanity (2010).

In short, deontological motivations have a passive rather than active character. However, when it comes to an environmentally directed act, the significance of deontological motivations lies in the fact that they form a value basis for it. Nevertheless, the effectiveness of the deontological approach in the environmental sphere is significantly diminished by the fact that society does not yet have a firmly established social norm relating to nature. At the same time, by relying on a duty to a social norm, the individual is not burdened by a sense of personal responsibility, is not overly attached to the result and does not have great expectations. Their attitude is resistant to failure and the potential for disappointment is low (Hirschman 2002).

Virtue ethics does not set a purpose or goal for specific behaviour either. Its individualistically motivated adherent asks: "What kind of person do I want to/should I be?", with the arbiter of good qualities being largely a deontological view.[55] An important feature of virtue ethics is the integrity of the personality, which consolidates good qualities and underpins the whole way of life. As virtue ethics is not attached to a specific action, it has no concrete expectations. For centuries it was cultivated by heroic, Aristotelian (MacIntyre 1981) and hedonistic virtues, which shaped a sceptical attitude: The individual cannot fundamentally influence reality.

Until recently, virtue ethics almost exclusively applied to interpersonal relations; in the past few decades a specific branch of virtue ethics known as *environmental virtue ethics* has taken shape, according to which an interest in the welfare of nature is part of a well-lived life (Sandler and Cafaro 2005; Treanor 2008). Even in the environmental version of virtue ethics, there is an emphasis on the inner integrity of its adherent. The great figures of environmentalism are held up as inspirational role models not only for their love of nature and faithfulness to it, their courage and perseverance in its defence, but also for their self-sacrifice, generosity, kindness and sensitivity towards people. It may seem surprising that benevolence is commonly ranked as one of the major features of environmental virtue ethics (Frasz 2005).[56]

Likewise, it may be unexpected given the stereotypical ideas about the nature of environmental attitudes that in spite of the focus on the well-being of nature, environmental virtue ethics are not biocentric. Binka (2008, pp. 96–99) would regard virtue ethics as a representative of weak anthropocentrism. It displays features of anthropocentrism in that it acknowledges the exceptionality and excellence of humans as a species (Sandler and Cafaro 2005), including their capacity for interspecies altruism, compassion for other creatures (Ferkany 2011) – a human quality which contrasts with the moral indifference of nature (Dillard 1974). Environmental virtue ethics thus place humankind in the role of steward or shepherd of other beings.[57]

In examining virtue ethics further, it is useful to distinguish two general concepts for a way of life which can be termed *Lebensführung* and *Lebenskunst*. This distinction is, in fact, identical to the difference between *intentional* and *voluntary* modesty, which is significant from an environmental standpoint.[58]

The term *Lebensführung* – in English perhaps 'life conduct' – was introduced to sociology by Max Weber. It denotes *value-rational action* which is not in the direct interests of the self. 'Heroic individuals' (Weber 1922a) create their personality indirectly, through their self-sacrificing 'service to a cause'. In conjunction with goal-driven teleological motivations, *Lebensführung* is part of civically engaged Aristotelian ethics oriented towards the well-being of the community.

[55] An example of the interconnection of ethical types.

[56] Nevertheless, through the influence of the media we have come to associate environmental virtues more with assertiveness and intransigence.

[57] We will return to the issue of stewardship and the figure of the shepherd in the theological key (Chapter II.5.).

[58] I wrote about intentional modesty and how it differs significantly from voluntary modesty in *The Half-Hearted and the Hesitant* (Librová 2003, pp. 27–31).

[59] Kees Schmidt writes of a feeling of elitism accompanying an eco-friendly lifestyle.

[60] One famous advocate of a postmodern variety of *Lebenskunst* (*savoir-vivre*) is Michel Foucault (1984). For more on the categories of *Lebensführung* and *Lebenskunst* from an environmental perspective, see Librová (2010).

In conjunction with deontologically grounded asceticism, it is part of Christian ethics (MacIntyre 1981). Viewed sociologically, *Lebensführung* finds its reward in the strengthening of identity and self-esteem and even a feeling of exclusivity, of belonging to an elite (Schmidt 1993).[59]

The environmentalist version of *Lebensführung* reflects and accentuates opposition to the mainstream way of life which harms nature: "I'm not going along with this (and I won't change my mind, even if it complicates my life)." In conjunction with teleological elements, this form of environmental virtues results in *intentional modesty* (Librová 2003, p. 28) and ecological activism; according to James Connelly (2006), environmental virtues are at the core of 'environmental citizenship'.

Lebenskunst,[60] the 'art of living', is devoid of asceticism and incorporates a moderate degree of physical and spiritual pleasure; it is reminiscent of the classical *ars vivendi* of the Cynics, Epicureans and Sceptics, the representatives of hedonistic philosophy. Ronald L. Sandler and Philip Cafaro (2005) emphasize that the lives of adherents of virtue ethics are enriched by joys which mainly come from the non-material sphere – for example, the aesthetic or spiritual and contemplative dimension.[OK]

The ethic of *Lebenskunst* is the basis for an everyday way of life which has been summed up as *voluntary modesty* (Sandler and Cafaro 2005). In environmental ideology this has been expressed by the slogan 'Simple in means, rich in ends' (*e.g.* Devall 1988). It is not a matter of principled decision-making: its adherent finds the majority's wasteful way of life unappealing. The individual does not see a more modest, environmentally friendly life as self-denial, but rather as something agreeable and desirable, even if there were no environmental problems. In the *Lebenskunst* lifestyle mode, concerns about the impact of the environmental crisis recede into the background and are almost forgotten. This approach is typically linked to artistic activities and an attraction towards spiritual matters. The weaknesses inherent within the *Lebenskunst*, in contrast to the strengths of teleology, are the inability to take expert knowledge fully into consideration, its intuitive and improvisational character, and at times the presence of a kind of ethical narcissism. These characteristics relate to the ephemeral, often simply faddish, environmentally friendly 'art of living'. Think, for example, of imported Far Eastern lifestyles or the so-called 'bourgeois bohemians' (Brooks 2000), which differ greatly from the enduring traditional deontological norms.

An important point for this book is the following: Even if an individual motivated by environmental virtue ethics embarks on activities with a specific goal, they are not overly attached to the outcome; in this they differ from an actor whose motivation is solely teleological. Whether we label this attitude as an Aristotelian or Stoic-like form of 'life conduct' (*Lebensführung*) or an Epicurean-style 'art of living' (*Lebenskunst*), the adherent of environmental virtues shares one trait with their ancient predecessors: They are aware of an external locus of control[PK] and harbour a greater or lesser degree of scepticism. As was said of the deontologically motivated attitude, if an adherent of environmental virtues has no great expectations, they are in no danger of being disappointed (Librová 2013).

Seminar students in the lecture hall.

Seminar study B

Students attending the 'Lifestyle and Environmental Issues' seminar carried out two research surveys in the spring semesters of 2014/2015 and 2015/2016. Survey B focused on identifying motivations for an environmentally friendly way of life. Students were to search for a person whose consumption patterns met the criteria for the Colourful sample from 1992. They presented quotations from interviews confirming that the theoretical motivational categories are readily transferable to the interview guide through operationalization. The results of the survey pointed to the likely predominance of *virtue ethics*, especially its *Lebenskunst* variant; its adherents have relatively little involvement in civic life, which is particularly striking in comparison with the Colourful.

The seminar discussion around the survey pointed to aspects of the way of life that are often neglected in studies on voluntary simplicity: firstly, the willingness to reduce productivity or working hours or modify them to make them more environmentally friendly (part of the phenomenon of 'downshifting'), and secondly some forms of recreation which may be no less environmentally devastating than excessive consumption. We looked at the hypothetical question of how the Colourful's way of life would evolve if it were to become widespread. Would it go the way of all fads?

In the seminar of spring 2016, we discussed the individual chapters of the forthcoming book. The discussion with students gave me a better insight into young people's way of life. I became aware of their sensitivity towards the suffering of animals and updated my lexicon with a few choice items.

II.5. Theological key:[TK] Steward, shepherd, sibling

The religious basis of environmentalism

After many years as a sociologist amongst environmentalists, and as a teacher amongst students of environmental studies, I have noticed a growth in interest in religion.[61] It has to be borne in mind, of course, that over the past few years or decades, atheism has been replaced by a kind of indistinct religiosity throughout society. People in late-modern society have remained sceptical and critical of religions, in particular of the Roman Catholic Church, though there still remains a tolerance and respect of religious faith. During the research into the Colourful, as in other studies it was revealed how important a person's spiritual dimension is for the creation of ecologically friendly ways of life. As a result, I have added a fifth interpretative key – the theological key.[TK] Due to the ambiguity of theological themes in environmental circles, we will look at this in more detail.

On a daily basis, environmentalists convince themselves of the seriousness and magnitude of ecological problems. As has been the case in history, people who are helpless and anxious have turned to authorities which are not of this world. They offer some kind of defence against unsettling events in everyday life[PK] and provide hope for the lives of their children.[62] However, for environmentalists, Christianity is not merely a useful tool which can reduce their frustration and perhaps get new allies. Without even being aware of it, environmentalists and Christians have strikingly similar characteristics.[63] These include trying to live a modest life and caring for the weak and the afflicted.

According to analyses by the American historian Thomas R. Dunlap, published in the book *Faith in Nature: Environmentalism as Religious Quest* (2004), environmentalism not only displays religious features but it has a religious essence. It is wrong to look at the basis of environmental movements in political terms; although environmentalists argue mainly about the interests of society, they and religiously minded people are, in effect, addressing the universe. Even if environmentalism emphasizes a communal character, it is based on the profound inner attitude of the individual, on personal faith. The attempt to live a 'green' lifestyle, to make everyday individual behaviour more ecologically friendly, is not only a defence strategy,[PK] it also derives from the search for the meaning of existence and one's place in the world.[OK] The consequence of such attitudes is a strong relationship between environmental activists and their fellows – similar to the one we might

[61] The founder of ecology, Ernst Haeckel, would have been horrified. During his lifetime, this prominent zoologist was famous for his anti-religious attitudes. He refused to accept that nature was the work of a Creator. He attempted to defend his passionate and militant atheist diatribes using his own slightly wilful interpretation of Darwin's theory of evolution.

[62] In chapters IV.1.–IV.3., I mention how some of the Colourful reacted to the question of how to cope with the consequences of the ecological crisis by stating without hesitation: "I am a Christian. My faith protects me."

[63] The subtle sociological sense for features of new Christian communities that are similar to environmental groups was shown empirically by Jan Váně (2012).

observe at a meeting of young Christians. According to Dunlap, environmentalists see the destruction of the wilderness as a sign of profound spiritual disharmony, they cast cautionary prophecies of doom, as well as promise heaven on earth if we mend our ways. Christianity and environmentalism lean towards conservatism in an unconscious symbiosis with the notion that things can change for the better through a kind of intuitive indeterminism.

Although it is possible to agree with Dunlap that environmentalism and religion have much in common, any attempts at mutual understanding between Christians and environmentalists have run up against obstacles. The central reason for Christians' caution and reservation in their attitude towards environmentalists is that they suspect them of 'neopaganism' (Komárek 2008, pp. 247–249).[64] In addition to Christianity, some American and West European movements have indeed been inspired by pantheistic religions – by the African and Native American indigenous rites, mysticism of pagan cults, the cult of Gaia, Mother Earth and by Buddhist traditions in particular (Lužný 2000). However, the actual influence of these religions within Czech environmental movements is weak.[65] If you hear the sound of African drums at the meetings of environmentalists, it would be probably just an exotic addition to the evening programme, sometimes with ironic intent. The listeners and drummers have no idea about sound's origin or meaning. I once attended a seminar of environmental education teachers at Želiv monastery and I was taken aback by the fact that the evening programme, held in a Christian building, included shamanic drumming – however, others were surprised what is my problem.

Until recently, one obstacle to cooperation between environmentalists and Christians was their opposite views on population size. In 1969 Gary Snyder wrote: "There are now too many human beings, and the problem is growing rapidly worse. (…) The goal would be half of the present world population, or less" (Snyder 1969). In his speech given at a meeting of Earth First! called "Grand Canyon Round River Rendezvous" in 1987, Dave Foreman said: "Even though we love this bunch of three or four hundred people here, though we love all the little wolves, there are still too damned many of us on the planet. Overpopulation is a problem" (Foreman 1992, p. 8). However, having a sizeable family[66] is still one of the basic teachings of Christianity.[67] Today, though, the anti-population aspect of environmental ideology has disappeared. The families of environmentalists tend to be fairly big. In chapters IV.1. to IV.3. we will discover that there is a relatively large

---

[64] Isn't 'neopaganism' a misleading negative label in connection to environmentalism? Is pantheism fundamentally incompatible with the concept of the greatness and beauty of nature as the work of a Creator? This question is addressed in the philosophical, religionist and theological texts of Leonardo Boff, Eugen Drewermann, Matthew Fox, Hans Küng, Karl Rahner, Henryk Skolimowski, Teilhard de Chardin and others. Czech writers include Bohuslav Blažek, Erazim Kohák, Dušan Lužný, Zdeněk Neubauer and Ivan O. Štampach.

[65] If some communities do have neopagan rituals, such as the celebration of the solstice and the full moon with the chanting of mantras, they have their source in social environments which are not linked to environmentalists. These include living-history groups (Vojtíšek 2007, p. 6), the New Age movement (Zbíral 2007) and the feminist movement (Reid and Rabinovitch 2008).

[66] The blood family was not always the central Christian value. In early Christian churches the basic unit was a group of fellow believers.

[67] It is remarkable how the theme of natality was ignored by Pope Francis's encyclical *Laudato si'*, however radical his idea of 'ecological conversion' may be. It was this indifference that may have fuelled strong reaction of journalists to Francis's statement at a press conference during his visit to the Philipines on 19 January 2015: "Some think that - excuse the word - that in order to be good Catholics we have to breed like rabbits" (Garsd 2015).

number of children in the families of the Colourful, and the lifestyle of the entire household largely revolves around them.

The objections to Protestant denominations of Christianity mainly stem from the Calvinist work ethic, according to which God's favour is granted through success in work. The environmentalists see the glorification of work performance, which transforms the natural world into useful products,[OK] as one of the causes of the ecological crisis.[68] Idleness, including a pleasant rest in the countryside, was viewed by Calvin as a grave sin.

The greatest obstacle in the relationship between Christians and environmentalists is often considered to be Christian *anthropocentrism*. This position, consistent with the idea of progress as the basic principle of history, has pre-Christian Judaic and ancient roots, and is deeply ingrained within Christianity. Until recently, anthropocentric views were given short shrift by environmentalists. Lynn White presented the widely referenced opinion that "Christianity is the most anthropocentric religion the world has seen" (1967, p. 1205), that it is based solely on the perspective of what is beneficial for man and for the prosperity of his descendants, and that this belief had significantly contributed towards the origin of the ecological crisis. This idea soon caught on and thus required revising. Thanks to theoretical studies of environmental ethics, education in environmental philosophy and gradual changes in theological research, this primitive view of anthropocentrism has been shown to lack foundation. It is in fact a Christian perspective which enhances environmental ethics as it elaborates on an anthropocentrism in its promising, mainstream 'weak version', which leaves mankind its central role but at the same time "considers environmental protection as one of the highest values" (Binka 2008, p. 99).

A privileged culprit

The foundation for Biblical anthropocentrism is to be found in the First Book of Moses (Genesis 1:26): "And God said: 'Let us make man in our likeness'". Other passages from the Bible show that amongst his creations, man is exalted and loved by God. It is little wonder, therefore, that environmentalists are uncomfortable: How is it possible that a creature made in God's image has destroyed beautiful landscapes and awesome forests, polluted life in the seas and disrupted the climate for himself and other creatures? It is not only the environmentalists who are bitter and horrified. The German theologian Jürgen Moltmann

[68] This aspect of Protestantism, or rather Calvinism, is examined by Lynn White (1967). I refer more to his ideas in *The Colourful and the Green*.

stated that "What we call the environmental crisis is not merely a crisis in the natural environment of human beings. It is nothing less than a crisis in human beings themselves. It is a crisis of life on this planet, a crisis so comprehensive and so irreversible that it cannot unjustly be described as apocalyptic. It is not a temporary crisis. As far as we can judge, it is the beginning of a life and death struggle for creation on this earth" (1993, xiii). Another Protestant theologian, Jiří Mrázek, who specializes in the last book of the Bible, The Revelation of John, sees clear signs of an ecological catastrophe in the horrors of the Apocalypse. However, he does not believe it is God's punishment for people's behaviour, but rather "a heavenly projection" of "what people *themselves have caused* (J.M.'s emphasis) through their behaviour" (2012, p. 110). And Pope Francis (2015) wrote: "Because of us, thousands of species will no longer give glory to God by their very existence, nor convey their message to us. We have no such right" (para. 33). "Never have we so hurt and mistreated our common home as we have in the last two hundred years" (para. 53).

It is characteristic of Christianity that not even difficult confessions of guilt need result in despair. Repentance is possible. If a culprit repents, there is hope in the form of 'ecological conversion'. "To achieve such reconciliation, we must examine our lives and acknowledge the ways in which we have harmed God's creation through our actions and our failure to act. We need to experience a conversion, or change of heart" (para. 218).

Theology, therefore, proposes for those who would destroy nature to mend their way. Interpreters of the Scriptures have identified two emblematic figures which express man's relationship to nature: *the figure of the steward* and *the figure of the shepherd*. In truth, preachers often fail to distinguish between the two – at one point the person is a steward of nature, the next he is its shepherd. This is a pity as these two symbols express fundamental differences and allow for more profound reflection.

Steward

According to contemporary theology, the relationship between humankind and nature is expressed in verses from the First Book of Moses: "The Lord God set man in the Garden of Eden to cultivate and guard it" (Genesis 2:15). As for other creatures, nature has been given to humanity as a source of sustenance, but the difference is the guardianship conferred on people, which is performed on the basis of free decision-making. Freedom of action is the basis of anthropocentrism and is self-evident and essential in both Judaism and Christianity. Sometimes we hear that in fact this is not anthropocentrism but rather theocentricism, because nature was created by and belongs to God, man is not the lord of the Earth, he has merely been placed there as a responsible steward.[69] However, the emphasis on his dominant position in the world is undeniable. In the vocabulary of environmental ethics, the concept of steward is strongly anthropocentric in character (Binka 2008, pp. 96–99).

As we shall see in Chapter III.2., a strong type of ecopragmatist has entered today's environmental scene. Ecopragmatists no longer try to save the wilderness; they themselves take control of the management of ecosystems and the planet Earth. They envisage technological solutions being applied to ecological problems and are not burdened by virtue ethics. Where are the limits for the steward-ecopragmatist as pace of civilization's technology blindly accelerates?

Even a divinely ordained steward sees and administers the world using human senses and tools, based on erroneous ideas. Hans Jonas writes of "man's precarious trust" (1987, p. 5). Evolutionary anthropologists and evolutionary psychologists admit that human cognition is limited to a certain degree; that humankind has a cognitive deficiency. The combination of evolutionary-based psychological characteristics with the increasing power of technology has proven to be ecologically dangerous. Cognition and behaviour are focused on a small socio-geographical circle within the simple logic of cause-effect. People are interested in the short-term perspective: their imagination about the future is restricted or flawed due to a tendency to extrapolate.[PK]

Can a steward with these characteristics administer an ingeniously created world? As I often mention in this book, natural processes are incredibly complex, interconnected and full of unforeseen consequences and unexpected setbacks. Where can a steward turn to for the criteria of his decision-making? Before abandoning his role and returning his task uncompleted to the Creator, as he is teleologically motivated[MK] he resorts to a proven source of knowledge – science. However, he finds little here too. Today, you hear more and more from scientists about how difficult it is to understand the world, that it is in fact unknowable as it responds inexplicably to human behaviour, even to the helping hand of its protectors. How then can a steward make sensible judgments concerning genetic manipulation, research into nanotechnology, alternative sources of energy or advances in astronautics?

[69] With characteristic sarcasm the famous physicist James Lovelock remarked: "Originally a steward was the keeper of the sty where the pigs lived" (1991, p. 45).

Hypothetically speaking, due to the state of today's world it would be environmentally appropriate if the steward were to put a stop to the rapidly accelerating dangerous inventions and behaviour.[70] "If the common catastrophe of human beings and the earth is still to be avertible at all, then it is certainly only by synchronizing human history with the history of nature, and if the experiment of modern times is carried out 'in accordance with nature' and not in opposition to nature, or at nature's expense. (...) It is essential to 'cool off' human history, and to slow down its one-sided varieties of progress," wrote Jürgen Moltmann (1993, pp. 137–138). Something as unimaginable and unprecedented in history goes against the mentality of today's wilful ecopragmatists.

Shepherd

It would appear that the concept of a rational and responsible anthropocentric stewardship is a dead end. Might it be possible to find someone else – mandated by the Creator – who is more suited to the task? If we were to keep to Biblical metaphors, we would find another anthropocentric one – the poetic image of the shepherd reflected in the figure of Jesus Christ. The theological suitability of this metaphor was shown in an issue of the journal *Universum* entitled "Man – the Shepherd of Creation" (Czech Christian Academy 1994).[71] Unlike rational, managerial, ecopragmatist purposefulness, this emphasizes spontaneous goodness, kindness and faithfulness.[UK]

The shepherd's kindly attitude, empathy, self-sacrifice towards weak neighbours and altruism in general are all deontologically based[MK] features of Christianity and would appear to be a good starting-point in the search for cooperation between environmentalists and Christians. The qualities of the good shepherd are similar to that branch of environmental ethics known as *environmental virtue ethics*[MK] (Sandler and Cafaro 2005). At its foundation is benevolence (Frasz 2005); friendliness, forbearance, an almost condescending attitude towards the weakness of nature,[72] and an altruistic outlook. It is a strange form of an anthropocentrically superior loving relationship and has been part of the philosophy and life of, for example, Albert Schweitzer (1990). Even today it is the dominant motivation of nature conservationists, in particular those who rescue individual animals in need.[73]

Altruism is often defined as a person's ability to risk or sacrifice their own wellbeing (sometimes even their life) to help others. A sceptic, especially if guided by a neo-Darwinian perspective, would doubt this.

[70] In an allusion to the slogan 'sustainable development', James Lovelock spoke of the need for 'sustainable retreat' (2006, p.7).

[71] However, it is remarkable and surprising for the reader that the motif of the shepherd is not developed in the text itself or even mentioned.

[72] I am aware of the argument that nature is strong rather than weak. However, the perspective depends on the position we are perceiving the situation from and, indeed, on the chronological scale.

[73] In contrast to ecosystem preservationists, whose motivation is more of a rational and systemic character.

[74] Note the special jargon of the Neo-Darwinists.

He would say that altruism does not in reality exist because "the donor of the altruistic act"[74] always expects that the attention will be returned to him in some way, if not in the form of reciprocal service, then in a token of gratitude or at least in the feeling of increased self-esteem.

Environmental altruism is much more complex than altruism towards people. The altruist who cares for nature can only seldom expect gratitude – perhaps when looking after an intelligent pet. Conservationists talk of the grateful looks from dogs after being freed from suffering, however, cats seem to be extremely aloof in the way they receive human attention, rabbits are not very communicative and it's highly doubtful we would recognize a chicken's expression of gratitude for being freed from a cage and moved to a courtyard. In any case, the central aim of environmentalism is not the protection of domestic animals but the maintenance of habitats, living conditions and the prosperity of animals. Apparently, this impersonal though environmentally fundamental type of altruism cannot be expected to become a source of gratitude in any respect.

And a reward in the form of increased self-esteem? We must remember that the consequences of environmental action tend to be felt far in the future, and whether it will work can be unclear and have no guarantees. Activists who manage to save wetland from the construction of a shopping mall have to accept that the result of their efforts might only be temporary and that in the foreseeable future it will be destroyed anyway. Where then can 'he who carries out the altruistic act' find satisfaction?[75] It would appear that environmental altruism is one of the rare cases of 'pure' altruism.

As it is hardly a means of earning money, an act of environmental altruism might be compensated by religion, by some kind of merit featured in the karmic religions. The idea that merit would be registered and valued in heaven held currency in the Christian Middle Ages and contradicts the Reformation notion of the relationship between man and God which rejects this 'meritorious' idea. Instead of this, Christianity had the example of Christ's sacrifice and an altruism which (in the language of sociobiology) was not conditional upon reciprocal behaviour. Unconditional altruism, contradicting the neo-Darwinian idea, therefore represents an exciting common basis for Christian and environmental thought and action. The environmental weakness in the Christian concept of the shepherd is to be found elsewhere, specifically concerning the issue of who the recipient of the altruistic act is. While nature only knows of intraspecies altruism – its creatures only look

[75] Professor of theology Jan Heller used to tell environmentalists to disregard the results of their work.

[76] Occasionally we hear that it is possible to observe interspecies altruism in nature. Upon closer inspection, however, it usually proves to be a complicated example of reciprocal or nepotistic altruism.

after themselves, their offspring and their kin[76] – the foundation of nature conservation is a culturally created form of interspecies altruism.

Who does Christian altruism apply to? Who is the neighbour that the Christian/shepherd can and wants to take loving care of? I harbour the suspicion that in the vast majority of cases the neighbour is only a fellow Christian in the spirit of strong anthropocentrism, regardless of the Gospel of John's part where we can read: "I have other sheep that are not of this fold" (John 10:16). The question arises of how a Christian understands "sheep from another fold". One interpretation of the Gospel verse which we hear in sermons is that these are Christians from other Christian denominations, more seldomly they are followers of other religions – the broad ecumene. We very rarely find a clergyman who understands them as non-human creatures. However, I did find a passage in the Old Testament which theological interpreters and preachers have missed, despite the fact that it could work well in sermons: "Man has mercy on his neighbours, but the Lord has mercy on all creatures" (Sirach 18:13).

The reason they have missed this might be simple and factual: Sirach is a deuterocanonical book which the Reformed churches, whose practices rely heavily on the interpretation of the Scriptures, do not have in their Bible. But another explanation also occurs to me: It is such a radical and extreme statement – reminiscent of Francis's encyclical – that theological interpreters have avoided it.

Across the world, for example in Great Britain, there are Christian groups whose work involves the direct protection of animals (see SARX 2015). One particular curiosity, containing symbolic meaning, is that in some foreign parishes domestic animals are invited to attend the service. In Český Brod, Father Matúš Kocian[77] invited churchgoers to bring a domestic animal with them to mass.

When I look around me, I see that many practising Christians have a loving and caring relationship with the animals in their households. But when I listen to zealots and those without such experiences, it seems to me that a position of strong anthropocentrism and intraspecies altruism persists and they have the suspicion that 'people are overdoing this love of nature' when they should be helping people in need around them.

Sibling

Since the 19th century there have been arguments between natural scientists about whether a clear boundary exists between *homo sapiens* and other species, and whether a hierarchical order exists in nature. They have asked in what way and to what extent evolutionary development has imprinted the traits of animal predecessors in humans. Prominent authors who have examined these questions include the ethologist Konrad Lorenz (1966), the sociobiologist Matt Ridley (1996) and the biologist Franz Wuketits (1998).

In recent years, voices doubting the strict division between humankind and nature have also been heard from humanities disciplines such as sociology, philosophy, culturology, anthropology and theology. The biologist and professor of theology Celia Deane-Drummond specializes in an area which had previously been thought of as an exclusively human domain: reason, freedom, morality, language and justice. She discovered that for certain animals, in particular primates, it was possible to observe behaviour which showed signs of culture and morality (2014). She claimed that the border sharply dividing humans from other creatures, which had formed a dualist, anthropocentric view of the world, had been artificially constructed. Whatever the case, according to the majority of anthropologists today, the border between humankind and other creatures is not sufficiently clear to justify ethical anthropocentrism.[78]

Naturally, anthropological research was not the intellectual inspiration for the *Laudato si'* encyclical, and it was more as a mystic than scholar that the pope found himself in unintentional agreement with evolutionary anthropologists. Francis understands the relationship between humankind and nature as a sibling relationship. There is none of the dominance contained in stewardship or the dominance which establishes the benevolence of the shepherd – it is a relationship where co-created creatures have a mutual origin and fate, a Buber relationship of "I and Thou" (Buber 1937). We find only one mention of stewardship in the encyclical in paragraph 116, which is not even an expression of Francis's opinion; it is merely a comment on another ecclesiastical document.[79] In paragraph 221, Francis explicitly encourages the establishment of a "sublime fraternity[80] with all creation". We repeatedly find the metaphor of creatures as "sisters" or "brothers" in the encyclical and in its key areas – in the closing hymns and in the well-known *Canticle of the Creatures* by St Francis of Assisi (para. 87):

Praised be you, my Lord, with all your creatures,
especially Sir Brother Sun,
who is the day

---

[78] In which direction is evolution headed anyway? Towards the most complex creature or the best adapted? If it is the best adapted, then according to evolutionary biologists humankind is not the favourite in evolution but certain types of insects.

[79] This is the declaration *Love for Creation. An Asian Response to the Ecological Crisis* (Federation of Asian Bishops' Conferences 1993).

[80] In evolutionary terms, humankind can be viewed as the younger aggressive brother.

and through whom you give us light.
And he is beautiful and radiant
with great splendour;
and bears a likeness of you, Most High.

Praised be you, my Lord,
through Sister Moon and the stars,
in heaven you formed them clear
and precious and beautiful.

Praised be you, my Lord,
through Brother Wind,
and through the air, cloudy and serene,
and every kind of weather
through whom you give sustenance
to your creatures.

Praised be you, my Lord, through Sister Water,
who is very useful and humble
and precious and chaste.

Praised be you, my Lord, through Brother Fire,
through whom you light the night,
and he is beautiful and playful
and robust and strong

From an environmental perspective, the weak anthropocentric comparison to siblings has a more significant impact than the metaphors of steward and shepherd, as it addresses a lifestyle which each person freely chooses each day. There is a word in German which precisely expresses this lifestyle – *schöpfungsgemäß* – which is appropriate to the vital needs of other beings that have been created in the world. If we weren't destroying creatures by the way in which we live, we wouldn't have to be technologically adept stewards in charge of the planet's ecosystems, or compassionate shepherds, intuitively protecting endangered individuals and biological species. The principle of siblingship, reflecting today's dire environmental situation, offers us the courage to turn to the metaphor of a helpless and suffering God and through this to the difficult theological question of theodicy. "It is not by his omnipotence that Christ helps us, but by his weakness and suffering," wrote the German Protestant theologian Dietrich Bonhoeffer from a Nazi prison on 16 July 1944 (1959, p. 164). The famous treatise *The Concept of God after Auschwitz* by the Jewish philosopher Hans Jonas (1987)[81] also questions the idea of an omnipotent God/magician who can solve the ecological crisis for us, thereby exposing the flimsy 'defence by faith'.[PK] According to Jonas, God is not all-powerful. The world he created was entrusted into human hands and he is descending into misery and suffering; he suffers along with every weak, helpless creature.[82] And thirdly, the Catholic concept of a suffering God, complete with an expressive urgency reminiscent of Spanish Baroque, is contained in Francis's encyclical: "Mary, the Mother who cared for Jesus, now cares with maternal affection and pain for this wounded world. Just as her pierced heart mourned the death of Jesus, so now she grieves for the sufferings of the crucified poor and for the creatures of this world laid waste by human power" (para. 241).

If we were to summarize the content of this chapter as: People were set down in the image of God – *imago Dei* – in what sense are we to imitate the qualities of God? The image of God as a wise creator might be understood as a steward's attempt to administer the Earth. The image of God as kind and loving is of a nature conservationist, the shepherd protecting creatures from "other sheepfolds". How can someone trying to protect nature be inspired by the suffering of a defenceless God? Is it insolent to see in environmental grief the siblinghood-based human imitation of a suffering God?[83]

---

[81] This was the ceremonial lecture given by Hans Jonas in 1984.

[82] Compare with the above-cited interpretation of the Apocalypse by Jiří Mrázek (2012) and with the principle of Cosmic Christ (Fox 1988).

[83] Chapter III.1. of this book.

Dietrich Bonhoeffer's prison letter to his parents
(taken from the book *Prisoner for God: Letters and Papers From Prison* [Bonhoeffer 1959, pp. 53–54])

"I have in front of me the gay bunch of dahlias you brought me yesterday. It is a reminder of the lovely hour I was allowed to have with you, and it also reminds me of the garden and of the loveliness of the world in general. One of Storm's verses I came across the other day seems to express my mood, and keeps going through my head like a tune one cannot get rid of:

And though the world outside be mad,
Christian or unchristian,
Yet the world, the beautiful world
Is utterly indestructible.

All I need to bring that home to me is a few autumn flowers, the view from my cell window, and half an hour's exercise in the courtyard, where the chestnuts and limes are looking lovely. But in the last resort, the world, for me at any rate, consists of those few we would like to see, and whose company we long to share... And if I could also hear a good sermon on Sundays – I often hear fragments of the chorales carried up here on the breeze – it would be better still."

13. 10. 1943

Dietrich Bonhoeffer was a German Protestant theologian. He was part of a conspiracy to assassinate Adolf Hitler and was arrested in April 1943. He was executed at Flossenbürg concentration camp just one month before the end of the war.

Faithfulness to nature and faithfulness to ideas

III.1. Environmental grief

[84] From an ecological perspective, Radek Štěpánek's observation is more precise:

"Touch-me-not has occupied another stretch of denuded riverbank, / (...) One by one the original inhabitants fall / into captivity, where they perish for lack of nutrients, air and sun / behind bars formed of fleshy stalks which know no mercy / and spread like wildfire until there is nothing left to burn."

(*Invasion*, Štěpánek 2018, p. 12)

You will never walk that sun-flooded path again
In July's clouds of the scent of meadows around old rivers,
You will never walk through the labyrinth of meanders
Ensnared by the river and the clouds
And smothered by the branches and leaves of maples and walnuts,
Hedge maples, white and balsam poplars,
You will never taste the scent of rivers again,
The fishy smell of the Dyje and the moldering mud of the Svratka
And will not see jungles of nettles with the hops and the clematis
Winding around wild and pink bells
Of Impatiens royeleyi...

This poetry, combining nostalgia with botanical knowledge, was used by the ecologist and photographer Miloš Spurný in his 1973 film *Sbohem, staré řeky* (Goodbye, Old Rivers), when unique floodplain forests disappeared below the surface of the Nové Mlýny reservoirs (Spurný 2007).[84] I have included Spurný's poem at the beginning of this chapter on environmental grief as a reminder of what I confessed to in the introduction: I feel dejected by the changes to nature which I have observed all of my life around me. I am not the only one among colleagues and friends who experiences and shares this environmental concern, but in our discussions we try to be cheerful and make light of the situation. I join in but I am left with rather uncertain feelings. It seems that by having a pleasant conversation and trying not to appear bitter, I am betraying something of fundamental importance. I have a feeling of being unfaithful towards, if not actually betraying, living creatures.[UK]

Surprisingly, it was through reading academic literature that I found my way again. I was relieved to discover that I had kindred spirits in the academic world: my fellow female scientists, thanatologists.

The opinion of biologists

In a review of a book by Jane Goodall, the well-known chimpanzee conservationist, the zoologist Jan Robovský wrote: "It is small wonder that I regularly meet colleagues, conservationists (most often in zoos), who feel either some sort of despair or despondency in their work" (2012). This is an accurate description and the parenthesis indicate something important: The closer an expert is in contact with

65

the fieldwork and with living beings, the more sceptical and sadder is their tone: "Studies focusing on the population size of common species [of insects – H.L.] also paint a terrible picture. (...) It is the picture of the clinical death of the remains of Europe's nature," wrote a group of entomologists in the journal *Vesmír* (Universe, Čížek *et al.* 2009). We can hear similar opinions from geobotanists, ornithologists and pedologists.

Researchers who make generalizations based on big data and render it harmless in models (optimally in evolutionary theories) are relatively well protected from the depressing impact of information on the adverse developments in nature. According to macroecologists, the ecological crisis has in fact an ambivalent nature due to the unpredictable complexity of natural and social processes. "Each collapse creates new opportunities, often wide-reaching," wrote the palaeobotanist Petr Pokorný (2008, p. 11) in his rebuttal of the pessimistic, old-fashioned nature conservationists. With the benefit of hindsight, evolutionary researchers can conform to the intellectual imperative of the late modernity, happily intellectualizing[PK] and relativizing: "To a large extent, whether it is good or bad depends on personal taste" (Storch 2008, p. 36).

However, as soon as macroecologists begin to research specific issues in the field, their opinions become more polarized: "Today woody encroachment and eutrophication [of backwaters – H.L.] are global and universal processes which are difficult to counter, but to say this is made worse by people farming in the countryside is reprehensible," said David Storch in an interview for the journal *Ochrana přírody* (Nature Conservation, Šůlová 2012, p. 33). And in an interview for the magazine *Sedmá generace*, he surprised as well as convinced readers, most of whom are passionate about wildlife conservation, with his opinion that woody encroachment of the Czech cultural landscape was one of the main factors behind biodiversity loss (Svobodová 2016).

It is interesting to follow the opinions which the world ecology heavyweights have on the state and development of nature. At the start of the 1990s, the famous zoologist and evolutionary biologist Edward O. Wilson came up with a rational solution to the environmental crisis: "The rescue of biological diversity can only be achieved by a skilful blend of science, capital investment, and government: science to blaze the path by research and development; capital investment to create sustainable markets; and government to promote the marriage of economic growth and conservation" (1992, p. 336). In 2006, in his book *The Creation* with the subtitle *An Appeal to Save Life on Earth,* Wilson started to view the situation for "the remainder of life" on the planet and the outlook for the Earth's ecosystems as so dramatic that he, an atheist, was forced to turn to a Baptist pastor with an urgent call for cooperation.

In his book from 1979, *Gaia: A New Look at Life on Earth,* the equally famous chemist James Lovelock acknowledged the seriousness of the ecological crisis but also wrote of the "fierce emotional drive" and "wild exaggeration" among environmentalists (2000, p. 108). And he also added soothingly: "A system as experienced as Gaia is unlikely to be easily disturbed" (p. 123). Some thirty years later and Lovelock's ideas have changed radically. He writes self-critically with an almost penitent tone: "I was so wrong. (...) The evidence coming in from the watchers around the world brings news of an imminent shift in our climate towards one that could easily be described as Hell: so hot, so deadly that only a handful of teeming billions would survive. We have made this appalling mess of the planet and mostly with rampant liberal good intentions. Even now, when the bell has started tolling to mark our ending, we still talk of sustainable development and renewable energy as if these feeble offerings would be accepted by Gaia as an appropriate and affordable sacrifice" (2006, pp. 147–148).

However, in one of his last works, *A Rough Ride to the Future*, Lovelock stepped back somewhat from his dramatic opinions and tragic language, though not from the fantastical nature of his solutions. These could involve, for example, symbiogenesis with technology which could lead to forms of life which are able to withstand higher temperatures than carbon-based life. Without either offering arguments or a more precise hypothesis, he writes encouragingly in the conclusion in the spirit of ecopragmatism: "Now is a critical moment in Gaia's history. It is a time of ending, but also a time of new beginnings" (2014, p. 163). The reader of Lovelock's book can't avoid the impression that the famous chemist is 'floundering around'. The books by Wilson and Lovelock reveal the arduous transformation in the ideas of prominent authors. We might pose a sobering question: Are Wilson's gesture towards the clergy and the undisguised misanthropy of James Lovelock mere signs of the intellectual deterioration of grumpy old men? Is it an expression of lifelong love for Gaia, which in spite of his efforts and those of others, Lovelock sees as damaged and sick at the start of the 21st century? We have to reluctantly admit an entirely different explanation: The possibility that the two experienced naturalists are not muddled old men, but that they have changed their outlook under the weight of information – which they receive more of and in advance of us.[85]

Deep ecology and thanatology

Despite their academic wording, there is an urgency in some books by naturalists today which is reminiscent of the texts from deep-ecology workshops, which in the 1970s and 1980s were a reflection of the powerless despair due to the suffering of animals and the extinction of biological species. The experts, however, would deny that such similarities exist. The methods of deep ecology, the rituals of evolutionary remembering which enable us "to hear within ourselves the sound of the Earth crying" (Seed 1988, p. 5), would appear to them as sentimental, uninformed, banal, bordering on bad taste and merely following fashion.

Certainly, in this last aspect they would be wrong. It is now thirty years on since the heyday of deep ecology, and the phrase 'environmental grief' has appeared in the academic literature, though biologists haven't had the courage to use this phrase yet. It was brought into the academic discourse by researchers from another field of science – thanatologists, or to be more precise, it was female thanatologists who were not afraid to use it.[86] They were undoubtedly reconciled to the fact that their texts were not going to be published in any high impact biology journals.

Thanatology looks at grief as a natural human reaction to loss. On a general level, grief is understood as the emotional anguish we experience when someone we love dies. Therefore, environmental grief is conditioned by a love for living creatures or for the countryside. The phrase 'environmental grief' was coined by an American, Kriss A. Kevorkian, executive director of the Center for Conscious Dying and Grieving, which operates from Antioch University in Los Angeles. She examined the phenomenon of environmental grief as a response to the devastation of ecosystems caused by human activities. For example, she investigated how members of the American Cetacean Society reacted to the death of the killer whale population in Puget Sound (2004).

Another thanatologist, Marie Eaton, director of the Palliative Care Institute at Western Washington University, wrote about the analogies between the stages of environmental grief and personal grief (2012), which she was familiar with through her work as a doctor. On an emotional level the individual at first feels powerlessness, anger and fury. Then on a cognitive level they resort to a denial of reality ('I don't believe it'). On a behavioural level environmental grief is manifested in paralysis and flight ('I don't want to see or hear anything anymore'). The fourth stage of grief consists of physical symptoms, typically of episodes of crying and tiredness. The psychologist John Fraser and his colleagues noted the characteristic expressions of personal grief amongst the environmentalists – the post-traumatic stress disorder (Fraser *et al.* 2013).

Kriss A. Kevorkian and Marie Eaton have repeatedly returned to the issue of how to approach the environmental grief of students, the existence of which they see on a daily basis. How are teachers to cope

[86] In academia it is almost exclusively women who write about environmental grief, and they do not hide the fact that they identify with it. According to an analysis carried out by Tereza Lehečková, more men than women publish in environmental journals. For example, in the popular journal *Resurgence & Ecologist* and in the academic journal *Environmental Values* the ratio of texts written by men and women is two to one. However, if we look at those articles which have explicitly emotional character (we chose the word 'love' in the title as an indicator in a content analysis), the ratio is reversed – there are more texts written by women and they lose the objectivity typical of academic texts written by men. One hypothesis which awaits a researcher is that we will not find environmental misanthropy in academic texts, fiction or poetry written by women. However, it is easy to find in texts written by men, for example in the aforementioned books by James Lovelock or in the poetry by Robinson Jeffers: *"It is easy to know the beauty of inhuman things, sea, storm and mountain; it is their soul and their meaning. Humanity has its lesser beauty, impure and painful; we have to harden our hearts to bear it"* (1965, p. 91). I have included an example of Czech Catholic priest and writer Jakub Deml's misanthropy in the following note.

with their own grief and that of students without losing the academic dimension and intellectual depth? How will their teaching colleagues view this method of education? How should teachers respond to the fact that there is no solution to the environmental crisis? (Weintrobe 2012, pp. 103–108).

Environmental grief can be understood psychologically and sociologically as the result of combining personal sensitivity with the interiorization of culturally induced nature's value. Marie Eaton sees the source of environmental grief in a different way. In an attempt to explain its roots, she refers to E.O. Wilson's biophilia hypothesis on the evolutionary connection between man and nature with its creatures. Eaton considers it evident that disrupting this bond leads to serious emotional anguish. It should also be mentioned that it were the deep ecologists, who unconsciously tended towards a biophilic interpretation of environmental grief when they practised "The Council of All Beings" in their workshops, involving evolutionary recollections: "Dive me deep, brother whale, in this time we have left. Deep in our mother ocean where once I swam, gilled and finned. The salt from those early seas still runs in my tears. Tears are too meagre now. Give me a song... a song for a sadness too vast for my heart, for a rage too wild for my throat" (Macy 1988, p. 75).

Senseless, culpable, unacknowledged grief
Although thanatologists understand environmental grief as an example of personal grief, there are also particular features which might make it especially urgent and intense.

The first feature: Death is natural in personal life, it is usually expected and it is unavoidable. Environmental loss is, therefore, so torturous because it is unnecessary and senseless. Species extinction contradicts our idealized view of 'eternal nature' filled with meaning. People who love nature experienced the absurdity of suffering when they reached the coast of Mexican Gulf following the Deepwater Horizon oil-rig accident in April 2010. Demetria Martinez and other volunteers tried to clean the marine creatures which were fighting for their lives under a layer of oil. She tried to help young turtles: "Then I did what I have worked hard to avoid as I've followed the coverage of the spill: I wept. The grief was unbearable as I gazed at the tiny creature, a wondrous manifestation of God's creation" (Martinez 2010).

The second feature: While the majority of us are not to blame for the death of those close to us, we can see environmental losses as the result of human activity or thoughtlessness. We are not innocent here. Some authors do not distinguish the level of guilt; they pass a general judgement that in our civilization we are all responsible for environmental losses. This feeling of guilt often leads to penitent, or at other times, misanthropic expressions of environmental grief.[87] This is a pre-war poem by Jakub Deml (2006, p. 110) which is curiously prescient:

We have sinned...

We have sinned, O Lord, before your eyes, we have denied Thee, like children,[88] who so easily believe in false hopes, we have remembered neither the past nor the present,
no longer like people, but like animals we have lived only for pleasure and all our desires,
and when our conscience chided us for throwing it all away, we said to ourselves that all power comes from the people
and that only we are lords of our body and our nation and we will create truth entirely anew
based on science and a completely new, rational religion
(...)

The third quality is essential: Environmental grief is abnormal and in fact deviant. This is why, unlike personal grief, human society finds it unacceptable.[89] Kevorkian uses the term "disenfranchised grief" for this state. The large number of people who experience environmental grief do not find sympathy, instead they are ostracized. This was illustrated by the ecologist Phyllis Windle when she wrote about how she sought in vain for an understanding reaction to her highly sensitive response to the news of the disease and death of a dogwood (1995).

Meanwhile, thanatologists stress that it is essential to communicate all types of grief, including an environmental one. The healthy response to grief is that it needs to be addressed. Pain cannot be suppressed – in isolation it paralyses the person, bringing anger and rage.

Environmental grief and 'green fatigue'
In the 1980s, environmental grief was shared by groups of likeminded deep ecologists, so there was a place to take refuge. In the rituals of 'despair and empowerment', or when reciting Robinson Jeffers and Gary Snyder, it was possible to convert grief and the feeling of powerlessness into renewed motivation for activism.

[87] "Admonish us again, O Lord, with the whip!" Jakub Deml called for punishment for the nation in 1938 (!) in a poem which is cited anon.

[88] Several sensitive authors highlight infantilism as a characteristic of today's society: "Could we grow up one day?" asks Radek Štěpánek (from his manuscripts).

[89] Modern society also tries to shut out personal grief, as is shown by sociological research into funerals (Ariès 1982).

"From the perspective of the other life-forms we will speak spontaneously among ourselves. We will say why we have come to the Council and be free to express our confusion, our grief and anger and fear. Then, after a while, to the signal of a drumbeat, five or six of us at a time will move to sit in the center of the circle to listen in silence as humans. (...) And lastly we will have the chance to offer to the humans (and receive as humans) the powers that are needed to stop the destruction of our world" (Fleming and Macy 1988, p. 82).

As was already mentioned in the introduction, at a time when the environment is coming under increasing threat, personal observation as well as research reveal a decline in the public's interest, which is linked to and strengthened by the media's lack of interest: there is talk of 'green fatigue'. Joanna Macy, who used to be a leading figure in the deep-ecology movement years ago, is eighty-seven. Although she is still active, the popularity of the workshops which she ran with John Seed has decreased. In place of the deep-ecology workshops which shared environmental grief, communities, which stress personal development and education on how to 'live lightly', are now forming. In the Rainbow Family's promotional material – perhaps the most prominent global alternative community – I searched in vain for the words of grief over the demise of the natural world. It would also appear that there are less direct actions to protect the environment, which the rituals of environmental grief used to be associated with.

In the words of thanatologists, environmental mourners remain misunderstood. Where could they seek understanding? Certainly not in everyday conversation – we've already explained why. There is one age-old retreat for mourners – in poetry. The biologists Pavel Kovář and Stanislav Komárek write excellent poetry. However, whenever they mention environmental grief, it is expressed only sparingly, usually in encrypted metaphors. Which poet of the weary late modernity has the courage to use strong, unambiguous words?⁹⁰ The *Almanach of Czech Poetry 2016* (Stibor *et al.* 2016) contains 270 poems; I found only one which contained traces of environmental grief.

Should we find refuge in older poetry which was not so afraid of banality? Today Deml's verse seems exaggerated to us, perhaps almost unbearable for some, Spurný's "you will never" reminds us of the aged nostalgia of Czech poet Antonín Sova. But their poetry contains the basic quality of environmental grief: they show with remarkable precision and knowledge what has happened to the countryside. In 1938

Jakub Deml was bemoaning the state of a landscape which we would view as an ecological paradise today.

> *...We have declared blasphemy and heresy*
> *as the only proof of honour and the good name for all,*
> *with no regrets for our children,*
> *whom we have thus sacrificed to the idols of*
> *Pretence and the Rotten Progress of our once healthy hearts!*
> *We have mocked our fathers and our ancestors,*
> *we established gyms and schools where it was proclaimed*
> *that only we could think and act and manage,*
> *we ploughed the boundaries between field and field,*
> *pretending that thistles, thorns and rose hips grow on them,*
> *which supposedly protect pests,*
> *all virgin land (on which our patriarchs grazed their flocks)*
> *like pigs we carved and drilled so that not even a butterfly*
> *could find a flower on which to rest and recuperate,*
> *we sprinkled every piece of bread with Chilean saltpetre*
> *and superphosphate,*
> *each boulder that preserved and transferred to us the wisdom of our*
> *ancestors,*
> *on the slopes, in the ravines and on the pastures, we shot into dust,*
> *from its greatness, from its beauty and from its whole*
> *we have reduced the faith of our fathers to thirty pieces of silver in*
> *the bank*
> *and crushed stone for the highways of democracy*
> *– It is bitter, Oh God, in our souls.*

(Deml 2006, pp. 110–111)

Optimism and pessimism; the cathartic role of sadness

This chapter looks at the intense environmental grief which sensitive people display in response to the suffering of living creatures, people who are exhausted and burned out by their long-term, futile efforts to protect nature. My academic friends and I[91] are usually only affected indirectly by environmental grief, but it is strong enough for us to be able to imagine how difficult it must be for Phyllis Windle and Demetria Martinez when working in their field.

Most environmentalists would not share such empathy. Some would even dispute the existence of environmental grief: It is necessary to

[91] And also the Colourful as we will see in Chapter IV.3.

encourage people and spread an optimistic mood within the movement and a positive agenda for the public (*e.g.* Kumar 2013). Direct opposition has even been formed with the provocative ideological label of "Dark Optimism" (Fabian 2014).[92] Its members are unhappy with the state of nature, but they vow not to succumb to paralysis and will act with "deep faith in human potential". "Dark optimism means being unashamedly positive about the kind of world we *could* create, but unashamedly realistic about how far we are from doing that right now" (ibid).

However, Hans Jonas considers it "absolutely necessary to free the demands of justice, goodness and sense from the lure of utopia (...) merciful scepticism stands against merciless optimism" (1997, p. 312). We might also recollect what the psychologist Marie Eaton said in an address to those who would spread optimism: "This kind of fear and despair cannot be banished by injections of optimist discussions of the technical fix or sermons on positive thinking" (2012, p. 9). Outside of the environmental sphere, when looking at the ethics bookcase, it would also appear that authors are not seeking optimism. In his study on existentialism, the literary historian and philosopher Václav Černý wrote bluntly on "stupid optimism" (1992, p. 52).[93] The conservative Roger Scruton was irritated by the optimistic character of interventionism, which would belong to our teleological category.[MK] He writes disparagingly about "unscrupulous optimists, whose goal-directed worldview recognises only obstacles but never constraints" (2010, p. 40). Against false hope it is necessary to admit that defeat is probable if not unavoidable. Paradoxically, though, it is only then that there will be a chance for change.

Despondent, deontologically[MK] based environmental grief has no objectives. It does not try to overcome obstacles and thus it protects the person from disappointment. Those who succumb to the pessimism of environmental grief expect nothing. But is such nihilism appropriate at a time when species are dying out at an incredible rate and when nature urgently needs protection?

Expecting nothing is not the same as doing nothing. In his previously cited essay, Václav Černý writes on existentialism that: "An action in the true meaning of the word starts from desperation. (...) Action in existentialism is anti-quietist" (1992, p. 53).

In psychotherapeutic practice, counsellors believe that an affective defence[PK] might have a cleansing and liberating effect, that environmental grief may have a *cathartic role*. Hamilton and Kasser (2009) tried to dispel the fears of educators that discussions on death and extinction lead to maladaptation or the materialist and consumerist behaviour of 'après moi le déluge'. If the theme of the end of life is not to be conceived of superficially, then communication and profound reflection on extinction and death will bring about a favourable change in behaviour. A psychological examination of grief might take us from maladaptive to adaptive coping strategies, to the sublimation and altruism[94] from our psychological key.

Therefore, the importance of grief lies in the fact that, due to the depth of the experience, it brings a permanence in attitudes – faithfulness[UK] – unlike the aforementioned indifference. Someone who is informed about and interested in the environment, but who is incapable of mourning or supresses grief, might end up at a dead end, leading to despair or to a chronic state of environmental melancholia and apathy (Lertzman 2015).

Grief and action, obstacles and scruples[95]

When we are in contact with environmentalists, we find hard-working activists afflicted by environmental grief to varying degrees. The majority have not suffered any acute attacks, rather they are grieving individuals 'cured' through contemplation;[OK] we are aware of it in some people, in others we can only guess. Chapter III.3. contains the results of my interviews with field conservationists.

It is understandable that environmental grief has become much more established in countries with a long tradition of nature conservation.[96] I will now focus more closely on the Dark Mountain Project, which originated in England as a response to intense environmental grief.

Paul Kingsnorth, a former deputy editor of the journal *The Ecologist*, had the courage to "accept the reality", as he himself said: "For fifteen years I have been an environmental campaigner and writer (...) I campaigned against climate change, deforestation, overfishing, landscape destruction, extinction and all the rest (...) But after a while, I stopped believing it" (Kingsnorth 2010). The words of his Canadian colleague, Dave Pollard, are even more dramatic: "I can't face another losing cause. (...) Even if we win, here and now, the developers will pop up like Hydras again and keep fighting until they eventually win. (...) We can get out in the streets and protest. (...) We can go on hunger strikes, or set ourselves afire. (...) We might in the process slow the development down for a few days, maybe even a few years. In the meantime

[94] A specialist in thanatology, Kenneth Doka, wrote about gender-related 'styles of grief'. Although gender is not a determining factor, it does have an influence, which is why Doka writes of the *intuitive* female model and the *instrumental* male model (Yalom 2010). As an example he presents a grieving couple who have lost the child they have longed for. Whilst the woman mourns day after day, the man works on a design for a tombstone. Naturally, this contradicts the image of the stoical woman we know from literary and theatre works, as well as the ideas of ecofeminism: it is the woman who takes matters into her own hands in times of strife and transforms grief into action.

[95] Václav Bělohradský writes of "the quiet voice of scruples" in a similar spirit (1991, p. 23).

[96] Czech society, lacking the traditions of romanticism, is not the best place for environmental grief to prosper. "Czechoslovakia painfully lacks the desperate and the foolish," wrote the poet Richard Weiner in 1929 in a review of Karel Čapek's *Tales from One Pocket* (Librová 2014b).

97 Pollard's website has an entire section (www.howtosavetheworld.ca) called Preparing for Civilization's End.

the people of the US, China and Canada will get their insatiable energy fix somewhere else" (Pollard 2012).

Where did these defeated individuals turn to next? Did they really find catharsis? Kingsnorth and Pollard believe it naïve to think that it is possible to avert the collapse of the model for contemporary civilization.[97] On 29 April 2010 Kingsnorth published an article for *The Guardian* website with the succinct title "Why I stopped believing in environmentalism and started the Dark Mountain Project" (Kingsnorth 2010). Instead of active resistance through direct action to protect the environment, participants in the Dark Mountain project decided to find refuge and solutions in an organized collective alternative culture, involving literary works, music, theatre, and crafts. They view the ecological crisis primarily as a crisis in culture; as a result they propose a radical cultural response which could bring about profound changes in society's values and behaviour. The people behind the project emphasize the collective character and social diversity of its participants, which is typical of a grassroots movement. In addition to former environmental activists, the project also brings together musicians, singers, designers, artists, engineers, teachers, artisans, gardeners and scientists.

Their manifesto (Kingsnorth and Hine 2009) for radical or even revolutionary cultural change seems at first glance to be teleological.[MK] However, it lacks the fundamental feature of teleology – the formulation of a concrete goal. The manifestos of disappointed and disconsolate activists have more in common with the virtue ethics mode of *Lebensführung*, which is deontological in spirit, based on proven traditions and rituals, artisan and agricultural skills; in fact, it does not contain any innovative features.

However, our question is: Is it possible to find catharsis in the Dark Mountain project and the transformation of environmental grief into action? If we were to be as strict with the sympathetic figure of Kingsnorth as we are in the next chapter with the ecopragmatists, we would hardly concede it was possible. However, relatively soon after establishing the Dark Mountain opposition project, which explicitly and spiritedly rejected active resistance in direct action, a piquant news story appeared: In an interview for *The New York Times* Paul Kingsnorth made a faux-pas when he admitted to the reporter that he had recently tried to stop a large supermarket from being built near his home in Ulverston in northern England. It is not easy for a leading light of environmental activism to dedicate himself exclusively to the establishment of a 'radically new culture'.

At the end of this chapter on environmental grief let's attempt to reflect on the findings which form the empirical basis of this book: the attitudes of the Colourful. Might they be an example of transforming environmental grief into action? I will offer references to the Colourful in Chapter IV.3., but I can give something away already:

In their case, it is not about the intense environmental grief of people burned out by their futile long-term efforts to protect nature; however, from our research interviews it transpired that the Colourful know about, or sense, environmental grief and it influences their daily behaviour. Their attitude towards life and nature fits Roger Scruton's description when he writes about "scrupulous people, who temper hope with a dose of pessimism, those who recognise constraints, not obstacles" (Scruton 2010, p. 40).[98] Nature does not need human actions in the form of intervention, and it wouldn't even require our protection[99] if people recognised constraints and adhered to them.

Naturally, this is all 'what if'. In reality nature today undoubtedly requires protection.

The poet retreats
(From Hana Librová's correspondence with Radek Štěpánek, 3 August 2016)

Dear Professor,
I am glad to hear that you like my poetry. In order to give you a more complete idea I am sending you the poems in their final form, my last two collections. Perhaps they not only contain environmental grief but also a little of the joy from the beauty all around... I've probably now resigned myself to the situation and I try to live life as it is and capture the world in its remaining beauty. Not that I am any less of a sceptic, but I realized that if I were to continue to be immersed in that scepticism, then my health would start to suffer!
Best regards

Radek Štěpánek

[98] In contrast to the "unscrupulous optimists" from above.

[99] Compare with the metaphorical character of the shepherd and sibling in Chapter II.5.

The astonishing zebra finch

Australian ornithologists discovered that some zebra finches (*Taeniopygia guttata*) sing to their eggs in hot weather. The newly hatched chicks grow more slowly and are more resistant to heat than chicks from eggs which their parents did not sing to – the smaller body mass allows them to cool more easily (Mariette and Buchanan 2016).

Song for an Egg

The sun can go no higher;
everything keeps its shadow to itself.
The garden stirs with a curious singing,
the harbinger of another heat wave.
In haystacks which they have carried
stalk by stalk up to the rusty
crown of the pine and in the crevices
of a crumbling stone wall,
zebra finches sing songs to their eggs.
Trills raised to the heightened pitch
of the rising mercury in the thermometers
flow over the shells. The Fates by the nests
pass on all they know about the trials
that are to come. Poised above the precipices of their
joys, they imprint their own life onto the embryos
so that it may once again search for eternity.
It must be this singing which allows
the unborn generation to survive.

(Štěpánek 2018, p. 7)

Ivan Zwach is an expert on amphibians and reptiles, author of the identification key.

Not even Mr Zwach's health problems would prevent him from carrying out his research work.

The extinction of a frog population behind a cinema
(from a research interview with Ivan Zwach – author of the amphibians and reptiles identification keys [Zwach 2009])

"I remember those night-time frog concerts. They came from all sides – you could hear those frogs everywhere. And when you consider that there were only two species – the common toad and the brown frog – and that only the brown frog called out in spring and with such a weak noise because it was underwater. (...) That population had to have been enormous to have been heard 150 metres away. From 150 metres you could hear the din those frogs made! You no longer hear that. (...) For example, when I started to write my first book, the area where I charted the behaviour of the brown frog was behind the cinema. At that time I counted some six hundred adults. If you go there today, you'll see five or six. They're all gone now."

III.2. Rational ecopragmatists

The ecopragmatist is a *homo faber*[OK]

[100] I will look at successes in nature conservation in Chapter III.3.

[101] I decided to use the word 'ecopragmatism' in this book. However, I will use the synonyms mentioned according to context.

[102] Michaela Kašperová offers an overview of the neo-environmentalists in her Master's thesis (2015).

The previous chapter focusing on the grief caused by the destruction of nature was based on an article published in the journal *Vesmír* (Librová 2014a). The reactions were interesting, yet not altogether surprising. Some readers wrote that the text spoke to them and finally expressed in words what they had been feeling for years. As I expected, others tried to put me right: It is also necessary to highlight what nature conservation has achieved. It seemed to me ridiculous and school-like to expect journalistic principle of balance in one essay article. Naturally, it is different for a whole book, where the author is indeed duty-bound to give a balanced view.

It is true that when we look closely and without bias, then we have to admit that some things have improved in the environment.[100] Teleologically[MK] based interventions of a technical nature are relatively quick and effective. Cleaner production processes have been introduced and you can buy energy-efficient domestic appliances. We have cleaner air and better water quality in some rivers thanks to intelligent 'green' technology designed by environmentally minded engineers.

The principles behind these ecologically friendly technological changes have been labelled in various ways, such as *ecopragmatism,* indicating objectivity, a focus on practicality and an overall rationality – *sensibleness.* Synonyms for ecopragmatism include *neo-environmentalism, post-environmentalism* and *ecomodernism*[101] – compound words which are supposed to indicate a radical and programmatic move away from the 'old' feelings-based methods of the 'old' environmentalists.[102] In 2004, the leading representatives of ecopragmatism, Michael Shellenberger and Ted Nordhaus, published an essay with the indicative title *The Death of Environmentalism.* The planet has been irreversibly scarred by human activity, the attempt to protect biological species is *foolish,* sentimental, old-fashioned and destined to fail. The sceptical advice of not wasting time on a lost cause has its justifications, and although we have no intention of being guided by it, we have to admit that it is quite rational.[UK]

One shared characteristic of ecopragmatists is their professional background, as they tend to be theoreticians who formulate argumentation frameworks for technological developments rather than as technologists themselves, as we would expect. The Breakthrough Institute, which is considered to be the think-tank of post-environmentalism, has attracted naturalists, biologists, physicists as well as economists and statisticians. Amongst the most prominent are the writer Steward Brand, the British journalist Mark Lynas, the statistician Bjørn Lomborg, and Adam Werbach, a former director of the prominent conservation organization the Sierra Club. The leading figure is Peter Kareiva, the vice-president and chief scientist of the leading American conservation organization The Nature Conservancy.

Ecopragmatists argue that we can no longer rely on the regenerative capabilities of a weakened nature. However, at other times they state that nature is not as weak as conservationists think and that it is surprisingly resilient. Whatever the case, people have to seize the initiative; rapidly disseminate scientific information and utilize new technology, nuclear energy, genetic engineering, nanotechnology, and geo-engineering.[103] In recent years, ecopragmatists have turned to synthetic biology.

A leading representative of ecopragmatism is the biology-trained journalist Emma Marris. Her oft-quoted book *Rambunctious Garden* (2013) contains the basic ideas of ecopragmatism: There is no sense in wasting energy and insisting on saving untouched and unspoiled environments as they no longer exist. The planet has been irreversibly transformed by human behaviour – for example, through climate change. It is necessary to take on the role of steward[TK] and create an environment which would be kind of a compromise between a nature in its wild form and an artificial one controlled by people. To the displeasure of conservative conservationists, Marris is an advocate of assisted migration, whereby species of plants and animals are moved to safer areas due to climate change or even personal taste. Is there any reason to suggest that the original ecosystem is better than an ecosystem designed by man? asks Marris. She believes that human activity can form novel ecosystems which will be the hope for the future.

Marris proposes acknowledging the importance of city parks and so-called urban wildernesses in industrial zones, as well as plots in courtyards which we no longer notice due to their ubiquity. It is indicative of her book and the texts by other ecopragmatists that they follow a strong anthropocentric line, focusing on the city, the purity of urban air, the health of people and the quality of food. There is no need to view things tragically – let's not be afraid of compromises. Readers are influenced by Emma Marris's openness, lack of prejudice, teleological spirit and optimism. She could serve as an example of 'positive thinking'.

[103] One typical example of geo-engineering is the proposal to cover the overheating Earth with an enormous sunshade. Less well-known is the 'iron fertilization experiment': the large-scale fertilization of the oceans with iron to encourage the growth of algae to extract CO_2 from the air. This experiment was carried out unsuccessfully – the researchers had to agree with sceptics that too many factors are involved in the processes in the ocean (Columbia University 2016).

However, we also have to ask ourselves the question whether rational ecopragmatism might not be an attempt to escape environmental grief or to transfer emotional affectation into purposeful rationalism. We investigated the personal histories of some of the most important ecopragmatists; these do not appear to support this assumption. It is difficult to imagine Demetria Martinez, distraught at the death of young turtles, maturing into a researcher in the field of 'assisted evolution'.[104]

The position of today's ecopragmatists is reminiscent of some of the ideas which the American novelist and journalist Michael Pollan already came up with three decades ago in his book *Second Nature* (1991). As with the ecopragmatically minded Emma Marris, Pollan believed that there no longer existed any areas in America where you could implement an exclusively protectionist solution. He considers the dichotomous choice – man or nature – as a nonstarter. He criticizes the metaphor of divine nature, as it allows for only two roles for humankind – that of divine worshipper or destroyer. Henry David Thoreau and John Muir "may have taught us how to worship nature, but [they – H.L.] didn't tell us how to live with her" (Pollan 1991, p. 189). Michael Pollan distances himself from the slogan of the radical movement Earth First! "No compromise in defense of Mother Earth!" and he doesn't hide his conviction about the suitability of an anthropocentric approach.[105] He criticizes the lack of faith in humankind and the radical conservationists' view that man's dabbling in spontaneous processes can only harm nature. According to Pollan, often "the only hope for the survival of another species is the manipulation of its natural habitat by man" (1991, p. 188). A welcome approach to nature is not the 'wilderness ethic' but the 'garden ethic', establishing loving human care.

However, there is a fundamental difference between the views of Pollan and today's committed ecopragmatists. Pollan wants the man/gardener to have a relationship with nature guided by a humility towards the *genius loci*. "The gardener cultivates wildness, but he does so carefully and respectfully, *in full recognition of its mystery* [italics H.L.]" (Pollan 1991, p. 192). Convinced ecopragmatists are not interested in respect for nature; in their programme they emphasize the servitude of nature, as we shall soon see. From a sociological and historical perspective, we can say that, unlike Pollan, the ecopragmatists broke with the old anthropological feeling of reverence and respect for nature as well as the cultural imperative of a love of nature[MK] – attitudes which were self-evident to environmentalists until recently.

The strikingly prophetic philosopher Hans Jonas would appear to have anticipated the future arguments of the ecopragmatists. In 1984 he wrote critically of "humanized nature". It is "a hypocritical phrase, embellishing its total submission to humankind with the goal of its complete exploitation for his needs. Because of this it must be radically transformed – humanized nature is nature alienated from itself" (Jonas 1997, p. 300). "Here humankind's love for nature cannot be enriched by anything, the riches and ingenuity of life cannot be taught here. Amazement, pious contemplation and curiosity lie dormant. The paradox is (...) that nature which is not changed or used by humankind is 'humane' as it is addressing man" (p. 302).

Pope Francis writes in a similar vein to the Jewish thinker Jonas: "We must be grateful for the praiseworthy efforts being made by scientists and engineers dedicated to finding solutions to man-made problems. But a sober look at our world shows that the degree of human intervention, often in the service of business interests and consumerism, is actually making our earth less rich and beautiful, ever more limited and grey, even as technological advances and consumer goods continue to abound limitlessly. We seem to think that we can substitute an irreplaceable and irretrievable beauty with something which we have created ourselves" (para. 34).[OK]

Each new 'ecological technology' is the result of the rational-creative activities of man the maker, to use Arendt's term. The ecopragmatist is a *homo faber*,[OK] who clearly defines his/her objective and task, and is driven to invent. This is based on the modern cult of the work-centered life (Arendt 1998, pp. 294–304), and thus enjoys the sympathy and trust of the public.

However, if we are interested in what the result of teleological solutions might be, then we begin to doubt the truly rational basis of post-environmentalism. *Homo faber* fails to take into proper consideration the unintended consequences of his/her actions. "Plato was the first to introduce the division between those who know and do not act and those who act and do not know," notes Arendt (1998, p. 223). As Henri Bergson said more concisely, "it takes us longer to change ourselves than to change our tools" (1911, p. 138). Ecopragmatists would not understand Hans Jonas's view on the necessity of fear and trembling.

Hardcore ecopragmatists are not *homines contemplantes*.[OK] They do not ask the fundamental question of virtue ethics "What kind of person do I want to be/should I be?"[MK] and neither do they concern themselves with Bergson's question concerning the essence of the life which they are intervening in. They are not interested in whether it is possible to fully understand nature and its responses to human intervention. There are ecopragmatists of the Pollan school who try to take into account the consequences of their actions, but even they are unable to estimate how nature will react – see Textbox "An attempt at creating steppe land" on page 86.

[106] The term 'Anthropocene' was suggested in 2000. However, some authors had previously looked at the idea of human dominance on the planet: In the 1920s the mineralogist V.I. Vernadskij coined the term 'biosphere'. At approximately the same time the palaeontologist Teilhard de Chardin used the terms 'hominization' and 'noosphere'.

Two concepts of ecopragmatism: 'Anthropocene' – 'ecosystem services'

A lexical analysis allows for a deeper understanding of latent meanings which even the user might not be aware of. We will look at common terms and try to decipher them based on context. Our discussion on ecopragmatism in this chapter can be expanded using two phrases – 'Anthropocene' and 'ecosystem services'.

The essence of ecopragmatism is aptly captured in the word 'Anthropocene'. The term was coined by the ecologist Eugene Stoermer and popularized by the atmospheric chemist Paul Crutzen, a recipient of the Noble Prize (Crutzen and Stoermer 2000).[106] From a geochronological perspective, the Anthropocene is the newest level of the Holocene. It's the period of human physical dominance on Earth, an era when people's activities have affected the planetary ecosystem in a fundamental manner. During the Anthropocene, the planet has been affected more by human activity than by natural processes. Since the beginning of the Anthropocene – between 1800 and 2000 – the human population has increased more than six-fold, the global economy fifty-fold and the use of energy by approximately forty-fold.

We do not usually hear the word 'Anthropocene' in everyday writings and speech – it is more the subject of expert and semi-expert discussion and argumentation. The participants are mainly theoretical biologists, physicists, geo-engineers and ecologists. While in previous decades futurologists had been disturbed by the depletion of the planet's resources , the main topic in today's debate on the Anthropocene has become climate change and the loss of biological diversity. As Clive Hamilton stated ironically (2014b), "a few social scientists and humanities people have been joining the fray, bringing their constructivist baggage," in order to cast doubt on the physical existence of the Anthropocene. In their eyes, it is a purely man-made construct.

However, the existence of the Anthropocene is obvious even to the naked eye. It can be seen in the changes to the landscape, large-scale deforestation, the redirection of river flows, opencast mining, transport construction and the domination of cities. Examples of the gigantic geomorphic impact of humankind include the Channel Tunnel and the Gotthard Tunnel; during their construction 20 and 24 million tons of rock respectively were excavated. A typical part of the Anthropocene is global neophytization (the introduction of new plants). Other processes of the Anthropocene are less visible to the layperson but are no less substantial, such as the changes to the geochemical cycle of elements including carbon, sulphur, silicone, phosphorus, nitrogen, mercury and heavy metals.

In order to understand the basis of the Anthropocene, it is important to look at the period when it originated. Natural scientists tend to link the start of this era with the industrial revolution in the 18th century, or even to the Neolithic period. Other academics, mainly social scientists, see the Anthropocene as being linked to capitalism and the hyper-consumerism which erupted after the Second World War and which led to the 'Great Acceleration'.

The term 'Anthropocene' is usually burdened with normative connotations. Environmentalists understand it negatively, with a bitter aftertaste. It reminds them of human expansion, arrogance and the domination of civilization, but also of their own defeat and powerlessness to protect nature. Ecopragmatists talk of the 'good Anthropocene' and understand it with a triumphal subtext. We can detect characteristically strong anthropocentric terms and phrases in the rhetoric of the advocates of the Anthropocene. In the Anthropocene, humankind is the 'manager' or 'steward'[TK] of a planet which provides him with 'services'. "Nature no longer runs the Earth. We do. It is our choice what happens here" (Lynas 2011, p. 12). "Environmentalists need to start seeing people as the solution, not the problem" (Werbach 2010). Some ecopragmatists elevate the principle of the Anthropocene to a level of visionary euphoria: they are convinced that humankind will leap to a higher level of planetary significance: "What a Great Time to Be Alive!" (Welch 2011).

Ecopragmatists respond to the concerns environmentalists have about the future of the Anthropocene by assuring them that 'mankind has always managed to solve his problems' or that 'technology solves everything'. Clive Hamilton and Tim Kasser critically point to neo-environmentalists' attempts at positive thinking at any cost: Coping strategies are a balm which might sooth momentary feelings of anxiety, fear and powerlessness,[PK] but they prevent any positive action (2009).

***

The second characteristic term for the ecopragmatic approach is *'ecosystem services'*. This probably needs no further introduction and the environmentally minded reader will easily understand why I am against this phrase, particularly when we try to transfer it into a verbal form. However, 'ecosystem services' became a notable concept in a significant United Nations project assigned by UN General Secretary Kofi Annan to scientists in 2000. This was a response to requests from several governments for information to supplement four already existing international agreements – the Convention on Biological Diversity, the Convention to Combat Desertification, the Convention on Wetlands, and the Convention on the Conservation of Migratory Species of Wild Animals. The basic question was: How do changes in ecosystem services influence human well-being?

The UN project was carried out between 2001 and 2005 by four working groups consisting of two thousand authors and reviewers from across the world. The Czech Republic was represented by the geochemist and ecologist Bedřich Moldan.[107] The project's summarized output – *Millennium Ecosystem Assessment: Ecosystems and Human Well-being* (Reid *et al.* 2005) – offers relatively precise data on the state of the planet based on the academic literature available. The project team was led by a committee made up of representatives from governments, businesses, nongovernmental organizations and indigenous peoples. The participation of these interest groups evidently had an influence on the content and wording of the outputs, which were in the spirit of ecopragmatism.

The Millennium Report contains both the strong and weak aspects of ecopragmatism. The strong elements include its clear questions, overall clarity and objectivity. It would be wrong to accuse the report of oversimplification – this was unavoidable due to the scale and objectives of the project. What is debatable, however, is the ideological or philosophical basis of the project. This is contained in the key phrase '*ecosystem services*'.[108] The report mentions provisioning, regulating, cultural and supporting services. While 'old' environmentalists are in awe of the beauty and ingenious ordering of nature and refuse to express its value in monetary terms, the ecopragmatic authors of this project viewed nature as a system of services for 'human well-being', whose 'marketed and nonmarketed benefits' can be quantified. The concept of 'ecosystem services' rationalizes and intellectualizes the gravity of the environmental crisis through exact methods.[PK] "Nature has its place in the human world principally as a provider of services," wrote Bedřich Moldan with latent bitterness in his monograph (2009, p. 344).[109]

Previous international conventions had been founded on biological arguments, and as a result they had shown a higher awareness of the essence and importance of natural processes.[110] The Millennium Report, which attempts to provide answers to a political assignment as well as respond to pressure from interest groups, ignores this perspective. The strongly anthropocentric concept of 'ecosystem services for the human well-being' became the implicit norm which neglected to protect nature for its own value. This is evident on every page of the text, in the formulation of the subthemes, and in the choice of analytical criteria. This fundamental flaw of the report is not altered by a three-line proclamation in the introduction, assuring us that the authors are aware of "the intrinsic value of species and ecosystems".

We do not have to be old-fashioned nature lovers to see 'nature's services' as being problematic, all we have to do is think objectively. One example is the category of 'supporting services'. According to the project authors, these include soil formation, photosynthesis, water cycling or the circulation of nutrients such as nitrogen. Every school pupil should know that these processes are the basic conditions for life on the planet and cannot be reduced to *support the well-being* of one biological species.

"It was a lot of work to plant and grow these cross gentians which one rare mountain Alcon blue feeds on."

An attempt at creating steppe land
(from a research interview with Professor Jiří Vácha; based on a text published in *The Half-Hearted and the Hesitant*, pp. 208–209)

JV: I wanted to buy land for butterflies...

HL: What led you to that?
JV: My uncle was a geography teacher. His bookshelves contained a large, illustrated book on the history of nature and books on butterflies and other things. I borrowed them and learned a lot from them. One of my aunts, who used to collect butterflies when she was young, got me interested in butterflies when I was ten. Then over the years I visited the reservations at Pouzdřany and Pálava. I knew that I wouldn't be able to buy one. I looked for something that might be turned into something similar. I naively thought that what I knew about forest-steppe from South Moravia could be simply reproduced by sowing what I'm supposed to and that would be it.

HL: So it didn't work out like that.
JV: The bush-tree part worked out – too well. I now have to cut it back because it's too thick. For the second year now I've been going there to cut wood.

HL: In an interview in 2002 you said that after two or three years, cornflower and sage began to appear, that two species from thirty survived.
JV: Perhaps even more after a time.

HL: In 2005 you said there was still hope. From twenty-five to thirty butterflies a day.
JV: That was the overall number I saw, not all at once.

HL: And now you're more sceptical.
JV: I am more sceptical. I'm more sceptical because I thought I could artificially increase the diversity here. There's a couple in Uherský Brod who are great supporters of butterflies and fauna. When they buy old pastureland, they don't have to do anything with it. They have everything there – thirty plant species. They bring me seeds from the White Carpathians, lots of different types that we no longer have here. In autumn they brought me wild gladioli, four types of seeds. When I realised that if I sowed them just anywhere and they never grew, I said to myself that I would pre-cultivate them. I bought boxes in autumn and it was a lot of work to plant and grow these cross gentians which one rare mountain Alcon blue feeds on. It lives in this area. I planted them and then I thought I'd see what'd happen. This year I can see how they're dying out. I'd say that the grass is still the winner. I can't stop the rain and make the meadow infertile – I know that now. I didn't know that before. In short, I'm very sceptical about what I can actively bring to the garden, but on the other hand, I have watched as nature itself has slowly increased this diversity.

HL: Is it fair to say you're worn out from all of this?
JV: It is, but that's obviously my age as well. These days if I go at this jungle for an hour and half with the brush cutter, then that's about my limit.

"I said to this one chap who has sheep: 'Let them graze wherever you like.'"

HL: And what if you were younger, knowing what you know now?

JV: It's hard to say what I'd do if I was 45, knowing what I do. As things stand, I'm happy with this forest park. At the time, a local agronomist told me: "There's enough fertilizer there for decades to come." She knew what she was talking about. In Brno they told me it would run out. But that nitrogen supply will apparently never run out...

HL: Nature hasn't exactly cooperated with you. Why have you persisted with the steppe land? Why didn't you make a grove?

JV: I need an open landscape, not woods. You wouldn't be able to see anything in a grove – you can't see into the distance. Here you have these open views.

HL: You're always talking nostalgically about gossamer-winged butterflies.

JV: Well, the butterfly in question lives in the old plum orchards near here. The gentian puts up with the grass there – I can't understand it, but it's always there. It's an old orchard that has dried up. I can't recreate that. I suppose there must be ants there. Why is it that it works there and not here? It's not exactly steppe there either. (...)

I find all that crowded vegetation so depressing that I'd rather strip the land bare than let it get overgrown. I said to this one chap who has sheep: "Let them graze wherever you like. Even on the rocks. Let the sheep trample the ground and eat what they like – it's better than all that grass." (...) My greatest enemy is nature itself, the grass and the bushes.

III.3. Un/faithful nature conservationists

Fatigue from strange activity

What actually is nature conservation? It is a strange activity: People try to protect nature from themselves, from their ever-increasing demands and the technology used to meet these demands. Nature conservation was established when society was guided by biophilic nostalgia and an instinct for self-preservation. However, the example of the protection of predators shows us that behind nature conservation also lies a contradictory motive – a cultural attempt to overcome the atavisms and old prejudices. Nature conservation arrived late in history and has remained marginalized within society. In spite of declarations in opinion polls, it actually goes against what people want, feel and value.

The 1970s and 1980s created favourable conditions for nature conservation and its institutions in Western societies. However, the successes of the environmental movement were relative and could not meaningfully counter the accelerating growth of technological civilization and people's increasing demands for material welfare. The influence of the environmental movement gradually weakened to such an extent that by the start of the 21st century, there was talk in the media of 'green fatigue'. A number of Western environmental activists were unable to hide their disappointment at their lack of success (*e.g.*, Kingsnorth 2010; Pollard 2012). The ecopragmatists Michael Shellenberger and Ted Nordhaus (2004) added fuel to the fire: They believe that the old environmental movements are unable to solve today's key problems; they describe the situation dramatically as the *"death of environmentalism"*. This negative atmosphere can be seen in society but also as a mood within the movement itself. We have seen the escalated form it can take – environmental grief. The world-famous environmental activist, Jonathon Porritt, stated that environmentalism was "too depressing" and "often too dowdy" (2007, p. 51). Some authors have argued that the dangers from 'green fatigue' may be even greater than the risks from the environmental crisis. In the 1990s, Joanna Macy and Molly Young Brown (1998) appropriately entitled one of their book chapters: "The Greatest Danger: Apatheia, The Deadening of Mind & Heart".

'Green fatigue' contains two mutually reinforcing related phenomena – *burnout* and *disappointment from failure*. Burnout syndrome is the result of the long-term expenditure of mental energy. However, when environmentalists use burnout as a metaphor for 'green fatigue' they do not mean in the sense of being exhausted by the intensity of their work. Instead, they understand burnout as the result of repeated failures, as the result of *disappointments*.[111] Unlike other *homines curantes*,[OK] there is a greater risk of burnout amongst environmentalists because disappointment occurs too frequently. This is due to the specific characteristics of environmental problems[UK] and people's negative attitudes towards nature conservation as a strange activity.

Conservationists' unfaithfulness?

In order to avoid burnout and transform environmental grief into action, you have to look at daily conservation efforts in context, expand the field of view, acquire distance and an overview – a mental process which social workers term 'existential analysis' (Längle 2003). The obstacles here lie in relativization and intellectualization,[PK] deconstruction and dilution, from which it is not far to unfaithfulness.[UK] Let us examine some cases of environmental unfaithfulness from real life.

One reliable way to avoid environmental burnout is to choose an ecopragmatic approach. Ecopragmatists do not deny the existence of serious global problems such as climate change, the shrinking of the ozone layer, the acidification of the oceans and the loss of biodiversity; but first and foremost, they are concerned about how this threatens the well-being of humans. They do not hide their unfaithfulness towards nature.

They openly advise moving away from the old-fashioned emotional conservation approach and no longer insist on saving the wilderness. According to them, natural ecosystems which have been untouched by humans no longer exist today anyway;[112] and even if they did, it is more sensible to forget the ineffective protection of these enclaves and turn our attention instead to cities as the human environment. And, of course, to support the development of new technology. Technological optimism leads ecopragmatists – *homines fabri* – to the belief that man could even recreate wildernesses if he so desired. The authors of *An Ecomodernist Manifesto* (Asufu-Adjaye *et al.* 2015) write about "the opportunity to re-wild and re-green the Earth" (p. 15). Apparently, such an opportunity will arise as soon as the inhabitants of the poor South reach the same material living standard as the richer countries – an opinion which every environmentalist with even an elementary schooling rejects.

The activists around the Dark Mountain project took a fundamentally different approach to the pragmatists in order to escape environmental burnout and the terrible state of the world – one which was

[111] The level of disappointment is based on the lack of success but also on the level of expectations (*cf.* Librová 2013).

[112] Climate change in itself significantly affects those areas where humans have not directly intervened.

[113] It is no coincidence that the magazine *Sedmá generace* has the subtitle "social-ecological magazine".

[114] The text gives an accurate description of the opinions of ecological economists and their critique of post-materialist theories, which view nature conservation as a luxury of rich societies (Inglehart 2008).

typical of modern *homines contemplantes*.[OK] In the chapter on environmental grief we read about the aesthetic and spiritual content of their festivals and about their inclination towards a new enchantment of the world. Critical readers are struck by the somewhat bizarre cultural codes of the 'green lifestyle' which were typical of the 1970s (Landy and Saler 2009). The first festival in 2010 near Llanghollen in Wales was advertised as a literary and musical weekend, which would also include the practical training of skills for an uncertain future "in a century of chaos" (Kingsnorth 2010). An 'arts and music festival' held in August 2012 at the parish of Hampshire Downs included a 'Funeral for a Lost Species', a celebration of the art of protest, storytelling around a fire, workshops on scything and foraging, a children's council, poetry readings, discussions and performances (Du Cann 2012). The organisers of these events, at other times burnt-out, desperate and extremely sceptical about the state of the world, spoke optimistically about the need for a profound change in the model for contemporary civilization and its values, and of the necessity to create a new culture. Is this the radical transformation of environmental grief into action, or is it rather an escape, an unfaithfulness towards nature in need based on cultural activities?

Amongst the opinions on solutions to environmental problems, the nostalgic cultural project Dark Mountain is something of a rarity. In the environmental theory and practice of some communities a current of a completely different character is gathering strength, which puts less emphasis on the prosperity of nature and is usually known as *ecological economics*. Environmental problems are to be solved systematically in conjunction with social and economic issues.[113] One aspect of this current is 'environmentalism of the poor'. At its foundation is the conviction that it is in the primary existential interest of the poor to protect nature as a source of livelihood. From the numerous texts, I selected a paper by the prominent representative of the social-ecological trend, J. Martinez-Alier, with the ironic title "The Environment as a Luxury Good or 'Too Poor to Be Green'?" (1995).[114] The idea of connecting social and environmental solutions is also part of the basic tenets of Pope Francis' encyclical *Laudato si'*: "The human environment and the natural environment deteriorate together; we cannot adequately combat environmental degradation unless we attend to causes related to human and social degradation. (...) For example, the depletion of fishing reserves especially hurts small fishing communities without the means to replace those resources; water pollution particularly affects the poor who cannot buy bottled water; and rises in the sea level mainly affect impoverished coastal populations who have nowhere else to go" (para. 48). Few people have engaged in a polemic with this opinion – and not because the author is the head of the Catholic Church – but because it is well-established within the influential left-wing current of environmentalism. Therefore, all the more interesting for the open-minded reader is the opinion of the conservative philosopher, Roger Scruton, who states that "social equilibrium and ecological equilibrium are not the same idea, and not necessarily in harmony" (2006, p. 34).

Either way, this case gave us another chance to observe how extraordinarily difficult it is to solve complex environmental problems. Over time, in alliance with the social-economic agenda, nature conservation starts to be seen as a burden and nature is sacrificed on the altar of economic prosperity and social and political stability. The economic situation of the poor is undoubtedly deplorable; however, it is eased by historical optimism, the generally shared prospect of 'better times' brought by economic and technological progress. Sober-minded environmentalists do not see an optimistic future; on the contrary, they believe that nature's capacity to mitigate environmental damage is reaching its limits. Here we see, therefore, that even with the promising case of an alliance between socially and economically oriented economists and environmentalists, faithfulness to nature disappears.

However, we don't only have to turn to the Global South and the probability that the poor who live there will abandon their current existential environmentalism as soon as they become richer – regardless of whether it will be under structural pressures or from the desire to have an easier life. A similar problem also exists for the environmental situation in our country. This is illustrated by an interview published in the magazine *Sedmá generace*. In an interview with the founder of the nongovernmental organization NESEHNUTÍ, which is involved in ecological as well as social issues, the question was asked: "The issue of human rights has begun to dominate the work of NESEHNUTÍ. Do you think environmentalism is dead?" The founder answered: "During its existence, the environmental movement registered no successes visible to the public. (...) However, the organizations dealing with human rights can offer a number of success stories" (Ander and Lužný 2004).[MK]

Let us now leave the world of academic reflections and look at what environmentalists do in their everyday work. It might be tempting to see them according to the old stereotype of the nature conservationist

Karel Zvářal prepares a nylon net for a 'dummy capture'. The owl is trapped by a model owl.

[115] This was a volunteers' initiative organized by the Klub ochrany dravců a sov. Those involved in Action FALCO protected the nests of birds of prey against theft as the chicks were being smuggled and sold to falconers abroad.

– an enthusiastic volunteer scything a colourful meadow, protecting a lynx family from poachers, saving the whales and ancient forests. The reader will easily guess that most of today's environmentalists are not to be found in the field looking after endangered species. To the surprise of the uninitiated, they can be mostly found sitting at their computers. They have become professionalized, they submit grant applications, fill out tables, prepare programmes for environmental training and education, contact journalists. Should we regret the fact that the environmental movement, which was established decades ago as activistic, has been tamed, and that in the words of Arendt, the caring person (*homo curans*)[OK] has been transformed merely into an active person (*homo agens*)? We will answer this question when we return from our walks in the South Moravian riparian forests and look at the case of the blockade at the Šumava National Park.

Faithful 'raptor devotees'

At the beginning of May I set off for South Moravia's riparian forest to meet some raptor conservationists. There were two reasons for this choice. These 'raptor devotees' are a distinct sociological and psychological type amongst ornithologists. The second reason was that I was able to use information I had gathered earlier in 1992, when I included a small survey amongst members of the Klub ochrany dravců a sov (Club for the Protection of Birds of Prey and Owls) as part of the research for The Colourful and the Green. In 1993 my student Marta Misíková (now Kotecká) provided information on the demographic and social structure of those who participated in Action FALCO[115] in her master's thesis (Misíková 1994).

Before Mr Hynek Matušík showed me the early-spring riparian landscape of Pohansko, we were sitting in a café in the town of Břeclav, where he told me about the conservation of birds of prey on the lower stretches of the Morava and Dyje. The conversation often turned to his ornithological discoveries of brown kites and red kites. In Kněžpole forest, near a cut-off meander of the River Morava, Karel Zvářal told me about observing birds of prey. It seemed to me that he had a particular fondness for eared owls, tawny owls, little owls and for "wee owls" in general. For our entire journey this uncompromising man, who wasn't afraid to criticize the authorities or professional ornithologists, was careful to make sure we didn't step on any snails.

The content of the field work of these 'raptor devotees' has changed over the past decades. In the 1980s and 1990s their main task was to

A registered, weighed and measured long eared owl ready to fly off into the moonlit night.

protect the nests from theft. Today there is hardly any need to guard the nests as scientific progress has made itself felt here too: A DNA analysis can identify chicks taken from the nests and convict the criminal. However, a new threat has surfaced; bait containing deadly poisonous carbofuran,[116] which is even more difficult to protect against. Today, in addition to helping track down the poisoners, the raptor conservationists' main work is ornithological research.[117]

Given the social dynamism of the past few decades, it is surprising that a club founded in 1990 is still active today, transformed into the Skupina pro ochranu výzkum dravců a sov (Group for Raptor and Owl Conservation and Research). These ornithologists have managed to avoid burnout. Some of those who I met through Marta Misíková's research in 1993 are still active, sacrificing not only all of their free time but also often their personal lives. There is a significant proportion of single men amongst raptor conservationists.[118] I learned from Mr Matušík how difficult it is to combine the role of husband and father with the role of raptor conservationist. "Either the wives suffer – sometimes along with the children – because they have to put up with the fact that you're travelling somewhere because a hierofalcon has a nest site there. And I didn't want that. Or there's a second option: Some of my colleagues abandoned the birds of prey to be with their families. I didn't want that either." The fact that holidays are sacrificed for the raptors is taken for granted. Raptor conservation seems to me more like a passion than a pastime.

In the universal key I wrote about faithfulness as an attitude in which compassion for the weak plays a role. Why do ornithologists not get together in similar exclusive permanent groups to protect larks, crested larks, wagtails – field birds, which today are among the most threatened groups of birds? Hynek Matušík agrees: "The waders, for

[116] I was surprised that this crime is spreading to South Moravia from Austria, a country with a strong environmental conscience. Austrian hunters are said to be a peculiar bunch characterized by the traditional 'k.k' (imperial.royal) access to the forest, including efforts to eliminate so-called 'vermin'.

[117] Moreover, the animal rescue station in Bartošovice tries to reintroduce birds of prey, in particular the golden eagle.

[118] Among fourteen raptor conservationists older than 25 who took part in Action FALCO, Marta Misíková found eight singles (1994). The research for *The Colourful and the Green* also found that single men were overrepresented amongst nature conservationists.

Equipment for ornithological work in the crowns of tall trees does not come cheap.

example, have disappeared here because we took away their environment – we drained and regulated everything. They've nowhere to nest, and if they do find somewhere, then we destroy it with our agricultural technology. There is a group which studies waders, such as lapwings and snipes, but they're not at the same level [as groups for birds of prey – H.L.]".

When I asked why ornithologists were so interested in birds of prey when, thanks to protection laws, their future is relatively bright, I first received a slightly school-level argument: "Birds of prey are at the top of the food pyramid." Karel Zvářal offered a chivalrous reason: "Until recently, birds of prey were persecuted." But at the same time, the two ornithologists speak with admiration about the impressive way in which raptors hunt. Hynek Matušík: "It's hard to explain. The research and conservation of birds of prey, as well as falconry, is a 'royal discipline'. It impresses people. It's about inaccessibility – how difficult it is to approach those birds of prey, how difficult it is to find their nests and climb up to those nests. It's quite an effort. At the age of sixty-four I climb up those thirty, thirty-five metres. I've got good equipment that I bought when I was earning something.[119] Just to get close to those birds, to their homes, is something that really amazes me. It's good that there's a group which looks after those waders and lapwings – but they are a class below." We can infer from the macho expressions of these raptor devotees that there are few women among them.[120] This confirmed for me what I had written in the universal key that faithfulness to nature is not only about compassion, and certainly not about self-denial; that it requires a fondness for the object of care. The admiration of hierofalcons, kites, falcons and eagles comes from their impressive size, strength, skilfulness and hunting elegance, their dominance among other creatures as well as their remoteness from people.

People's admiration of these qualities led to the establishment of a cult of birds of prey in ancient history, as can be seen in the famous portrait of a falconer on a Great Moravian coin found in Staré Město near Uherské Hradiště. Boria Sax examined the role which the cult of birds of prey had in the ideology of the Third Reich (2000). Today the interest in protecting raptors and large predators[121] can be seen as part of wider attempts at wilderness restoration (Monbiot 2014; Wuerthner *et al.* 2014).[122]

While talking with Mr Zvářal in the middle of a riparian forest lush with spring greenery and white flowering wild garlic, the singing of leaf and Sylvia warblers was interrupted by a rather unromantic topic: dealing with the authorities. Karel Zvářal evidently considers them to be important for his ornithological work, which was why he was able to fly off the handle: "Usually I was unsuccessful – it was a waste of time and just wishful thinking... For example, once I went to the police when someone shot a female hawk in her nest. They gave me a ticking off, saying that they were here to look for some businessmen who'd disappeared, and I'd dared to come and bother them because of some bird... And then those administrators act in a similar way, which is why I agree that their number and powers should be limited... They have the leverage to restrict the greed of the agricultural lobby, but they refuse to do it."

[121] Ornithologists, and zoologists in general, have found that in terms of biodiversity large biological species are relatively successful. This is not just the result of conservation but is also due to the size and structure of their brains, their intelligence and ability to adapt to different conditions.

[122] The so-called "rewilding of Europe" has two currents: efforts to protect the remains of untouched nature and attempts to artificially reintroduce wild animals (Bláha 2013).

 Hana Chalupská suggested that organisers themselves distribute a short survey right at the blockade. Another source of information on the structure and motivation of the blockaders was a survey done by the Department of Environmental Studies, Masaryk University, carried out one and a half years later (Pelikán and Librová 2015).

Hynek Matušík described his encounter with the authorities with a sense of humour as well as some understanding for the officials: "I found a poisoned sea eagle near Strážnice. I contacted the authorities and they sent the police and I waited with them until they had documented the eagle. They phoned me a fortnight later to sign a report. They spoke to me behind a glass screen, at a hatch... I'm a real nuisance to them cos I'm spoiling the police officer's performance record by finding poisoned bodies."

I was pleased to hear that ornithologists sometimes see negotiating[OK] with the authorities as being positive and perhaps even hopeful for the future: "The group that studies the waders has been talking to the Ministry of the Environment and the Ministry of Agriculture about some projects where farmers who leave the waders with land to nest are compensated." Hynek Matušík approvingly adds something which is the point of this whole chapter: "Maybe trying to stop that farmer from ploughing where the waders nest is more difficult and more necessary than me trying to protect the nest of a bird of prey."

The blockade at the Šumava National Park – evidence of faithfulness or rebellion?

Sceptical observers who doubted that environmental activists belonged in the camp of the faithful were to be surprised in the summer of 2011: Hnutí DUHA, the Czech branch of the international environmental organization Friends of the Earth, announced they were to carry out direct action in July – the blockade at the Šumava National Park. Journalists called it "a bark beetle war". In the tabloid newspaper *Blesk*, readers discovered on 27 July that "Activists chained themselves to trees!", and the day after "They attached themselves to trees using tubing!", on 3 August the newspaper *Právo* wrote that "Environmental activists in the Šumava protected trees with their own bodies".

We would expect the media image of the blockade and the blockaders to be superficial. However, it transpired that they had got things fundamentally wrong; in their estimate of the age structure, in their assumption that the blockaders were students, but principally in their motivations. Two open-ended surveys[123] showed that people of various ages met at the blockade – between 22 and 66 years old; importantly, the majority of people were over 30, and people in their 40s and 50s were strongly represented. As might be expected, the majority of blockaders were university educated or had just graduated, mainly in biological and environmental disciplines. But there were also

graduates and students from the humanities, arts and technical disciplines. This was consistent with the diversity in professions: those who
participated at the Šumava blockade included a computer scientist,
a doctor, a clerk, a low-voltage electrician, a deputy mayor from a Prague
municipal district, an ornithologist, an eco-counsellor, a doctorand in
biophysics, a Franciscan novice, an English teacher, and a pensioner.

By mid-August 2011, the blockade had ended; it was clear that it had
only slowed down the logging. In the area around Ptačí potok, most of
the trees had been cut down over approximately 80 acres of forest, which
had been in its natural state beforehand. Within three weeks, more than
10% of ecologically important waterlogged spruce forest had been destroyed. The main reason behind the organisation of the blockade was
now gone. We would have expected the blockaders to have dispersed
with a feeling of bitterness and futility.[124] The feeling of satisfaction only
came later when a ruling was made by the courts in Klatovy and Plzeň.
The blockaders and their sympathizers received more good news at the
end of May 2016: The Constitutional Court stated that the General Inspec-
torate of Security Forces had made an error by failing to thoroughly inves-
tigate Jan Skalík's criminal report and downplaying the police intervention against the blockaders. Despite having to wait a considerable length
of time, these court victories provided some moral satisfaction.

Did 'green fatigue' amongst the blockaders appear after the event
was over? After a year and a half, we asked them: "When now, with
hindsight, you look back on your participation in last year's Šumava
blockade (and possibly also consider the subsequent developments
at the location where the blockade took place, the judicial aftermath,
etc.), would you take part in the blockade again?" With the exception
of the oldest blockader,[125] the answer was a unanimous: "YES, I would
take part in another blockade". Even if we are cautious about this survey and acknowledge that disappointed and disconsolate participants
probably didn't respond, it is still incredible to witness such unity,
which is unique in attitude surveys. How are we to explain such resistance to failure and 'green fatigue'? Why are the blockaders from the
summer of 2011 demonstrating resistance by unambiguously showing
a willingness to return to the Šumava, while as we read in Chapter III.1.,
seasoned Western activists are giving up with a hint of desperation?
There might be a whole host of reasons and they'll undoubtedly be
interlinked. At first, there is the factual reason: The aforementioned
Western activists have been working tirelessly as leading figures in the
environmental movement, and now after a lengthy period they are

[124] The logging around Ptačí potok was not the first major failure Czech environmental activism suffered. For those who wish to recount some of the losses, examples would include the destruction of the village of Libkovice due to coal mining, a highway pass-ing through the Czech Central High-lands, and the futile resistance to the construction of the Temelín nuclear power plant.

[125] She apologised that she couldn't participate again for health reasons, which ruled out being part of a block-ade due to the harsh conditions.

exhausted and suffering from burnout. For the majority of the people at the Šumava blockade, however, it was a new experience. It wasn't another case in a long list of draining, failed attempts to protect nature, so there can be no risk of burnout.

Our analysis showed that virtue ethics predominated in the blockaders' motivation[126] in three fundamental aspects that are presented in the theory:[MK] an emphasis on personal development, benevolent attitudes, and political commitment. One blockader, an employee from Vodafone, wrote: "You test your limits, confront an extreme situation and thus get to know yourself." And a doctorand from a natural-science faculty: "Yes, I would participate because the experience cannot be conveyed or communicated, it is incredibly intense and thus very formative." Some respondents said that the blockade had been an opportunity to meet like-minded people, sometimes even as "fun": "It was fun too. It was incredible – that feeling of belonging and being committed to a worthwhile common cause," said a twenty-seven-year-old eco-counsellor, while a thirty-five-year-old mathematician/computer scientist added: "It was fun! I made a good friend there. We have been getting together ever since."

The benevolence which is characteristic of environmental virtue ethics is incompatible with the misanthropy present in certain currents of deep ecology. In our surveys we found no displays of hostility towards the loggers or local people. In fact, one respondent even expressed understanding for the police officers: "On a personal level, I understood why the police officers would be fed up as it was a Sisyphean task – to be constantly taking the blockaders away to the police headquarters and then they'd immediately go back again. I didn't mind them throwing me in the mud and trying to make me scared, but had I been in their place, I would have minded feeling like a puppet."[127]

The third feature of virtue ethics (that of the *Lebensführung* type) is commitment to community, so-called environmental citizenship. Although the objective of both the Shadow Scientific Council of the Šumava National Park and the organisers and spokespeople of the blockade was to reach "zero management" of the forest, and even though we could expect such teleological motivation[MK] from the ecologically and environmentally educated blockaders, the research surveys proved to be surprising in that the approach of the ordinary participants was different: Their main objective was not to protect centuries-old spruces. Their effort was not primarily about being faithful to nature, to use the terminology of this book. One characteristic survey response was: "We live in a democratic state based on the rule of law, but the absolute disrespect of the law and the absurdity of the situation forced me to express my opposition, to take time off work and allow myself to be yelled at by an emergency police unit. Nature is worth it." Some blockaders openly relativized the conservation reasons for motives based on general civic attitudes,[128] as was expressed by a thirty-year-old maths teacher: "I see the event more as a public protest against the misuse of power (the National Park director, the police) than as an attempt to protect a particular area of the woods. In the context of widespread corruption and social lethargy, I felt a civic duty to support a campaign coordinated in this way, a physical (not just a clicktivist) one... There are very few of them." Or the thirty-year-old urban planner: "I was not primarily concerned about protecting the trees, which had been cut down anyway, but about the unprecedented abuse of power."

Necessary *homines agentes*

When the newspapers published dramatic photos from the Šumava blockade, we were pleased that brave activists devoted to nature had chained themselves to ancient spruces to save them from the chainsaws. But in the survey they unromantically admitted that "it wasn't primarily about saving those trees". Other blockaders understood their involvement as "a public protest against the abuse of power, as political negotiating."[OK]

When I visited the raptor conservationists, I received a similar surprise: In order to observe kites nesting at the top of 30m-high oak trees, Hynek Matušík had sacrificed all of his free time and savings to buy expensive climbing equipment. But at the same time, he admitted that it might be better for the birds of prey if he communicated more with the officials from Forests of the Czech Republic about when and where to limit logging.

I became even more strongly aware of two aspects of conservation work from an interview with Josefa Volfová, the coordinator of a Hnutí DUHA project – Wolf and Lynx Patrols – which seeks to protect large carnivores in the Jeseníky, Beskydy and Šumava mountains. The annual report on the patrols' activities is more sociological than zoological. If the patrols want to effectively protect lynxes, wolves and bears, they have to spend a lot of time and energy communicating with the public.[129] They organize discussions with local communities, advise sheep farmers on how to protect their flocks, they lend electric fences and provide information on compensation for any losses incurred. They try to establish contact with hunters and persuade authorities and

[128] It is sociologically and psychologically significant that some of the blockaders at the Šumava forest were later present on the island of Lesbos as activists helping refugees (Hana Chalupská's personal comment).

[129] Certainly I'm not thinking of fundraising here.

citizens to join in the fight against poachers. They organize educational programmes and trips for children, and publish popularizing articles and leaflets.

The public's attitude towards these environmental efforts is at best ambivalent. The maintenance of wellspring boxes, planting of trees and cleaning away of rubbish in the forests are all activities which the public can easily understand, respect and appreciate. Nature conservation, however, often lies in administrative and political decision-making. *Homo curans* finds himself in the role of *homo agens*.[OK] Negotiating environmental causes is often very complex and confusing for the public. In addition, what might appear as a success to an environmentalist can be seen by other people as a step back in terms of their comfort and well-being. Let us remind ourselves of the initial characteristic of nature conservation as an activity which society enacted in an effort to limit the environmentally damaging short-term interests of institutions and individuals.

The *homines agentes'* fundamental objective is to promote citizens' environmentally friendly interests and monitor compliance with laws and regulations. They highlight mistakes made by authorities with 'rubber stamps' and uncover erroneous conclusions made by experts. And above all, in the interests of nature, *homines agentes* take the initiative themselves. We have read, for example, how they attempted to enforce the territorial protection of endangered species of birds of prey in negotiations with local foresters, or at other times the protection of endangered waders at a ministerial level. On a societal level, they lobby on behalf of nature's interests in the legislatures.

The course and outcome of negotiations are dependent on various circumstances. They depend on professional knowledge, the values and willingness of officials, but mainly on how prepared our *homo agens* is. They have to have experience 'in the field'; it would be difficult to be successful in negotiations if they only see nature conservation as a way of earning money while 'avoiding unpleasant fieldwork'. Part of the environmentalist's arsenal includes patience, the willingness to compromise and proceed tactically. At other times it is the ability to stand up and be counted in situations where they are perceived as stubborn 'green extremists' by officials, citizens and journalists. There are some excellent examples of representatives who link work in the field with negotiating with officials, such as our Slovak colleague, the wily and successful *homo agens* Juro Lukáč; the imaginative Brno politician and representative of the Green Party Mojmír Vlašín, and the expert on the Šumava forests Jaromír Bláha, who is also respected by his opponents. I have sympathy for the inventive ways in which the organization Arnika protects old avenues and defies the road lobby; I admire Miroslav Patrik's resistance – the famous 'nuisance' protecting the Czech Central Highlands against one and all; and the professional and systematic work of Marian Páleník, who protects the ecologically and aesthetically valuable riverbanks of the Elbe against the construction of weirs. I'm happy to say that I could go on listing names.

A cursory glance at the offices of environmental institutions might make us suspicious, but I believe that the *homines agentes* sat at these computers fulfil our basic criterion of 'enjoying fieldwork'. This makes them faithful *homines curantes.*

What has worked, what to look out for, what has to be kept in mind
(some advice for wetland bird conservationists from Vlastimil Peřina, director of the regional department of the Nature Conservation Agency of the Czech Republic, East Bohemia)

* If one winter doesn't come, it doesn't mean that the next one will (it is much easier to clear away mud during freezing winters).
* All permits must be granted for a period of at least three years longer than the original purpose.
* Fishermen know the ponds best.
* Areas disturbed by machinery which will not be below the surface can be very quickly overrun by bulrush, whose seeds are ubiquitous in the landscape.
* If the surface layer (even a thin one) with the rhizomes of the reed is removed and subsequently flooded with water, the reed does not come up - it "cannot breathe".
* Of fundamental importance is a carefully planned project. When preparing it, consult with the largest number of people who know the pond. Don't forget botanists and ice-skaters.
* Always photo-document everything carefully before and after.
* Mud islands are overgrown with willows in the first year.
* Pools can be created on islands and islands can be created within the pools on the islands.
* Be at the site as often as possible.
* Make friends with the diggers.
* That fact that you are convinced that you've explained everything in detail to the diggers and they will do it exactly the right way does not mean this will actually happen.
* Within half a year there can be sands with plover nests on the site of fifty-year-old alder forest in the middle of a peat meadow.
* The preparatory phases are time consuming and gruelling. Never-ending bureaucratic processes. There are often situations which look simply hopeless. Music might help here.

(Peřina 2015)

III.4. Evil nature. The anti-nature subculture

Two subcultures

To summarize what was said in the universal key on faithfulness to nature, we might say it is conditioned by our conviction that nature is good. Even though nature is something which generations of our ancestors struggled with like a formidable foe, its perfection and beauty have always dazzled humankind; European culture latterly forgot about some of the hardships and adopted the view that nature was good. However, if we set aside our rose-tinted spectacles that have been influenced by thousands of years of literary and artistic works and take a sober look at what is really happening out in the field, then we start to have our doubts. If we admit that nature is full of pain and horror, that it is cruel, then some unpleasant questions begin to surface which could influence our relationship. It is only once we have faced these questions that we can be sure that our faithfulness to nature is not blind, that it is not mere atavism or a false path that someone has set us on.[130]

In order to be clear: This chapter is not about the views of the eco-pragmatists we met in Chapter III.2.; they do not dwell on existential and ethical issues, they do not contemplate evil in nature. After all, ecopragmatists do not have a hostile relationship towards nature; only a superficial one which has become oversimplified through the constant search for utility. The anti-nature subculture more or less defines itself explicitly in relation to nature and ponders disturbing matters of a fundamentally existential nature.

However, we will begin by looking at some manifestations of the anti-nature attitude. When negotiating environmental cases which involve the participation of people from various professions – for example, in the negotiations concerning weirs on the Elbe, motorway routes, or the Danube–Oder–Elbe canal – it is interesting to observe how two hostile camps are always formed: the environmentalists and nature conservationists on the one side, and the proponents of technological progress on the other. For some reason, they are unable to come to an agreement, even in the cases where good will exists. Neither does it help to have an intermediary who has been trained in deliberative techniques. The two groups have different vocabularies and even, it would appear, different ears. During the pauses in negotiations, we can hear the disconsolate sentence in the corridor: "It's impossible. They're of a different mindset."

[130] The Czech writer Karolina Světlá rejected the duty to love nature: "I am only interested in mankind, this wretched, wretched mankind." In a letter to her friend Eliška Krásnohorská on 11 July 1857, she wrote: "By what coincidence did so much feeling go into this atom [the human heart – H.L.], while worlds circle around the sun in stupid, satisfied, bloated regularity?" (both in Soukup 2013, p. 134).

These core words indicate the existence of two subcultures which differ significantly in their relationship towards nature, and which in European culture do indeed have a 'different mindset'. Put simply, using the method of ideal types,[131] there exists a subculture of nature devotees and a subculture of their opponents, the anti-naturalists.

I do not have to explain in detail to readers the values held by nature lovers; I would only remind us of a few which are in obvious contrast to the views of the anti-naturalists. Rousseau-styled environmentalists argue that the main reason for the problems and dangers inherent in the modern age is 'alienation from nature'. The more consistent of them try, with considerable effort, to cling to nature through their lifestyles. However, in the 21st century it is not easy to transfer admiration for peasant or artisan craftsmanship into everyday life. And so we pay lip service to it by working in gardens or, if possible, community allotments. Environmentalists grow kohlrabi and carrots, weave baskets and make tie-dye T-shirts. For them it is important to be rooted in a place, in a specific landscape, but also in a home in the narrower sense of the word, in a family. We can easily observe environmentally friendly ways of life in popular magazines such as the British *Resurgence & Ecologist*.

The anti-nature subculture also has a characteristic lifestyle and particular aesthetic. One famous example is the lifestyle of the dandies in the 19th century. In his book *Sartre: The Philosopher of the Twentieth Century* (2003), Bernard-Henri Lévy describes one exponent of this subculture with a resolute, aestheticizing anti-naturalism and extreme anthropocentrism. He writes: "From the point of view of ontology, he [Sartre – H.L.] doesn't for a single minute believe in all those fairy-tales about nature" (p. 246).[132] For Sartre, nature is "the contrary of the dizzying, captivating enigma of Heidegger. (...) Nature is not lovable but odious. It isn't mysterious, it's disquieting" (pp. 246–247). Sartre had a morbid fear of animals which he did not hide. He looked on with horror not only at external nature but also at nature within man. He was disgusted by the body and the way its system works.[133]

Sartre demonstrated his distaste of nature and natural things with an almost laughable ferocity. He preferred tinned food to freshly prepared food, and meat with sauce rather than cooked 'naturally'. Bernard-Henri Lévy also mentions the painter Piet Mondrian, who had similar attitude. He would swap chairs in a café whenever he sat somewhere by mistake with a view of a garden or a tree.

[131] The method of ideal types attempts to describe certain characteristics in an extreme form; it consciously ignores their frequency and distribution in the population.

[132] At the time of Sartre's popularity, the anti-civilization movement flourished with its green emotional drive.

[133] Such resentment also expressed itself in his relationship with women. Sartre was excited when his lovers were clothed. He invoked an inner affinity with Charles Baudelaire, with whom he shared a penchant for actresses, women in make-up, and frigid and infertile women.

[134] I can't resist: "There is no soundness in them, whom aught of Thy creation displeaseth," wrote St Augustine (Augustine 2012, p. 158).

[135] Outside of the environmental context is an important sociological study by Norbert Elias. The "civilizing process" consists in replacing simple natural forms of behaviour with complex methods (2000).

[136] Today this idea is embodied in genetic manipulation.

[137] There is an inspirational collection for nature-loving readers who are happy to have their ideas challenged entitled *Is Nature Ever Evil?* (Drees 2003). It consists of essays adapted from contributions presented at a conference in Amsterdam in 2000. The issue of the presence of evil in nature is examined by philosophers, historians, theologians, biologists and other natural scientists.

Jean-Paul Sartre did not like what we would call home, "...what he never liked in the very principle of owning the house, was the way it was somewhere your being was deposited, a sedimentation of identity and existence. (...) Conversely, Sartre liked hotels and hotel life. He liked the anonymity of hotel rooms" (Lévy 2003, p. 230). "Sartre was an urban philosopher. He was the philosopher *par excellence* of tarmac and cafés" (p. 248). Perhaps even more important is a difference of an essentially ethical nature, even if it might appear less striking at first sight: Despite the ridicule from postmodernists, environmental activists declare they try to live a genuine and responsible life, while the café-society, anti-naturalist Sartre saw life as a game.

If we move outside the circle of our fellow environmentalists, we can observe what Bernard-Henri Lévy describes as the explicit and pronounced features of Sartre's philosophy in a less conspicuous and distinctive form around us. We can think of the lifestyles of some of our friends who have nothing against nature as such – they merely prefer days spent in cafés to roaming in the countryside. Sometimes we sense from coffee-house intellectuals a certain pose, a feeling of aesthetic superiority, an ostentatious unwillingness to succumb to nature-loving sentimentality. A Voltaire-styled disposition is opposed to the 'love of nature' sentiment, which had become almost obligatory since the age of Romanticism. At other times it is mere provocation or improvisation depending on the situation. When friends of the writer Egon Erwin Kisch invited him on a trip to the country, he refused: "A Jew belongs in a café." Our archaeology colleague refused to stick her head out of her cottage in the Beskydy mountains, saying that she wasn't interested in "birds and their annoying squawking."[134]

Evil nature

Representatives of the anti-nature subculture see alienation from nature as a positive feature of humankind and the foundation of culture and civilization.[135] And if we do have to live with nature, then it is necessary to improve it (Brooke 2003).[136] As can be seen from these examples, the antagonism of Sartre's followers towards nature has a marked aesthetic character. However, it also has a significant ontological justification linked to the issue of evil.

The philosophical debate on the idea of the presence of evil in nature goes back to the 18th century, an era which admired nature as the work of the Creator.[137] Historians see the earthquake in Lisbon in 1755 as its cause (Sanides-Kohlrausch 2003). The physician and famous obstetrician William Smellie did not share the conviction of his peers of a God bowing down with his love for living beings. He asked the question: "Why has Nature established a system so cruel? Why did she render it necessary that one animal could not live without the destruction of another?" (Smellie 1836, p. 222).

The ambivalent character of nature, sophisticatedly operating through its cruel food chains, has always disturbed many sensitive and thoughtful souls. "What kind of world is this, anyway? Why not make fewer barnacle larvae and give them a decent chance? Are we dealing in life, or in death? (...) The universe that suckled us is a monster that does not care if we live or die – it does not care if it itself grinds to halt," wrote the American writer Annie Dillard (1974, pp. 176, 179), appalled by the absurd extravagance of nature and its indifference to suffering.

The biologist and theologian Celia Deane-Drummond, who examines the theological implications of evolution, summarized that: Alongside 'moral evil', resulting from man's free will, there is also 'natural evil' (2008, pp. 114–116). This not only applies to natural catastrophes such as volcanic eruptions and earthquakes; it also has the character of predation and the 'evolutionary evil'.[138]

The world's leading evolutionary biologist, David Hull of Northwestern University, has written about evolutionary evil: "The evolutionary process is rife with happenstance, contingency, incredible waste, death, pain and horror. (...) The God of the Galápagos[139] is careless, wasteful, indifferent, almost diabolical" (1991, p. 486).

I am made aware of the conflict between the concept of nature in the subculture of environmentalists and in the anti-nature subculture when my students ask me what I think about vegetarianism. I answer briefly and somewhat jokingly that I honour vegetarianism, and even more so veganism, as an attempt to reach a higher humanity. And as a succinct caricature of this opinion, I would recommend a bold book by the Olomouc philosopher Marek Petrů *Možnosti transgrese* (The Possibilities of Transgression, 2005). Petrů is concerned by the same things as those vegetarians who are fighting against factory farming. Both he and they are motivated by compassion for farm animals.[MK] However, they are setting out from different ontological starting points and use contradictory arguments.

Environmentalists view factory farming as the result of humankind's alienation from nature; they try to prove that in evolutionary terms, as a biological creature, humans were not meat-eaters and vegetarianism respects our herbivorous heritage.[UK] If they are not vegetarians, they

[138] This theme has become part of theological reflections on 'theodicy', the issue of the vindication of God due to the presence of evil in the world he created (Deane-Drummond 2014).

[139] Charles Darwin carried out his observations on the Galápagos Islands, which are in the eastern half of the Pacific Ocean.

try to support organic animal breeding. Petrů turns the solution on its head: He emphasizes the need to consistently distance ourselves from nature. He points to the fact that the basis of intensive animal farming is a "biological nightmare". The rationalized mass-production of food – "the production of life in order to destroy it, while curtailing all of its other functions" (2005, p. 137) – only escalates this process "towards monstrosity". "Let us reject those scholars who try to persuade us that we are slaves to our nature, that it is impossible to rise above the biological," writes Marek Petrů (p. 243).

He wants to create a more moral and aesthetic post-biological existence. "By breaking out of the order of things, humankind reaches new levels – levels of humanity and ethics. Only then, when humankind has the ability to deny its own interests, will people become ethical and legal beings" (Petrů 2005, p. 133). "Humankind uses all means to transgress life, all means to seek to connect to the desperate emptiness of biological life – to the biological cycle of reproductive processes – to some kind of why, some kind of sense and meaning" (p. 135).

Thinkers from the natural sciences and philosophy wrack their brains futilely over such deep existential questions. Are theologists any more convincing when they find the meaning of biological life, including the horrors of predation and evolution, in the celebration of the Creator's greatness?[TK] "Faith allows us to interpret the meaning and the mysterious beauty of what is unfolding" (Francis 2015, para. 79). Is such a simple answer satisfactory to such a difficult question?

The ecological footprint of the Sartreans
We might expect anti-naturalists to live a life which is inconsiderate towards nature. But how large actually is the ecological footprint they leave behind? It doesn't necessarily have to be all that big. Those people who are 'alienated' from nature and declare themselves to be enemies of nature might even have a smaller ecological footprint than nature and countryside lovers. If, like Jean-Paul Sartre, you are holding court in a café, then you don't need a helicopter to go on an expedition to the Himalayas, you don't tread on the growth of orchids and you don't disturb nesting lapwings. If you live in a city and the countryside doesn't interest you, then it's easier to get around without a car. If you are happy living in a housing estate surfing the net, then you don't think about building a cottage in a protected area. You have no desire to see the last remaining exotic landscapes. In contrast, over the past decades, such a desire also afflicted many biologists and

ecologists. They are well aware of what there is left to see on the planet, and if environmental-education teachers managed to attract all of the 'alienated' to these areas too, they would lose their charm and nature itself would suffer.

If there is anything to be concerned about in terms of ecological footprints, then it is not from the private lifestyles of Sartreans. The most environmentally damaging behaviour comes from people's working lives, when they are drawn into the mechanisms of the economy.[140] Many people who would personally be inclined towards environmental protection are inadvertently forced by their work to damage nature. They do not want to, or cannot, leave their job and so they legitimize it by saying that nature is a hostile or alien world – a typical defence mechanism of reaction formation.[PK]

Representatives of anti-nature professions include architects and urban planners. It could be argued that the very essence of their work lies in taking over space and forcing out nature.[141] In addition, a distinctive, deliberate anti-naturalist subculture of architects has arisen[142] which rejects 'green' attempts to incorporate greenery into urban buildings.[143] 'Parks and trees have no place in a city,' say the radical urban planners, referring ostentatiously to the design of medieval cities.[144]

Professional anti-naturalism is even more evident among engineers – the builders of huge technical works which I mentioned in the ontological key. While an ecologist sees flourishing steppe or an area teeming with wetland life as being valuable, the engineer – *homo faber* – tells the journalist in front of the cameras 'there's nothing here now,'[145] until the plan is implemented and then there will be the perfectly executed multi-lane motorway, a dam, or an airfield.[OK]

[140] Václav Bělohradský talks of an "insane society of rational individuals" (2013, p. 188).

[141] So-called attempts at 'ecological architecture' do little to change this.

[142] Petr Hurník had a prominent position here, a popular teacher who will be remembered by a generation of architects.

[143] On the other hand, many architects such as the functionalists place an emphasis on greening cities. However, there is still a fundamental distinction between an artificially created environment and a unique natural ecosystem.

[144] It is a fact that having people concentrated in compact towns frees up space for life in the open countryside.

[145] The local citizen would probably add: "It's just a mess there" or more often: "It's a bloody eyesore."

Anti-nature IT workers?

Vojtěch Pelikán

[146] The group's heterogeneity would have to be taken into account when carrying out more detailed empirical research.

[147] A peculiar type of collecting containers ('geocaches'), which are often located in remote places and sought out with the help of GPS.

[148] Nevertheless, the listed activities might have environmentally unfriendly features.

It might be thought that professionals from informatics would have a position towards nature which is antagonistic or, at best, indifferent. The typical IT worker is seen as a pale being, spending most of his time by the flickering light of his screen. In general, environmental authors have tended to view information technology negatively and with suspicion (Louv 2012; Mander 1991). Psychologists and educationalists among them have warned that working with computers not only takes the individual away from nature physically, but it also alienates them mentally. IT workers are seen as the embodiment of these negative trends.

Who should actually be classified amongst IT workers? The Czech Statistical Office has a thorough catalogue of professions and divides IT into two main branches: IT specialists and IT technicians. The image of the IT worker is represented more in the first group – the better qualified programmer, developer or network administrator – than the more technically oriented 'maintenance worker'. This group in itself represents a very broad church. Not only are IT specialists from diverse social backgrounds with diverse political beliefs, but they also differ in their motivation. Some are attracted to the profession by a thirst for knowledge and a romantic DIY ethos, which gained momentum during the computer boom of the 1970s and which is symbolized by figures such as the founders of Apple, Steve Jobs and Steve Wozniak. An equally powerful motivation, however, is the prospect of a well-paid job.[146]

A closer sociological look at the group of IT specialists shows that initial judgment was misplaced. Firstly, it cannot be said that they avoid nature. Their hobbies include outdoor activities – jogging, larping and geocaching.[147] In their free time, many of them take their backpacks and head for the mountains. Even more important are the indications that this interest in nature goes beyond mere mental hygiene and compensation for time spent in a virtual world.[148]

We can even come across IT workers in research into environmentally friendly lifestyles. Their families figure among the so-called *amenity migrants,* who leave the city in an effort to find a better quality of life. IT professionals' households also appear in the research on voluntarily simplicity (Maniates 2002; Walther and Sandlin 2013). Moreover, we find numerous indicators of environmental sensitivity in books dealing with the lifestyles of workers in the creative industry, which to a certain extent interlinks with the IT world (Brooks 2000; Ray and Anderson 2000). IT workers are also well represented amongst owners of passive homes and members of eco-communities. They have advantageous conditions for such a lifestyle – their job is relatively well paid[149] and flexible in terms of time and location; it could be said they can live anywhere that has an internet connection.

There are also other findings, which have provided counter-intuitive insights into the relationship between IT specialists and nature. IT specialists often follow economic practices that are valued by environmentalists. For example, they are among the active participants in the *sharing economy* (Schor *et al.* 2016), which goes against the individualist zeitgeist. For example, in 2014 the Brno car-sharing cooperative Autonapůl carried out research amongst its members which showed that IT professionals made up one third of the membership. They also have a strong presence amongst supporters of environmental NGOs. The link between IT professionals and environmental issues even has an important political dimension: the Pirate parties, whose origin is in the free-spirited culture of programmers and hackers, place significant emphasis on environmental policies; on a European level they have made an alliance with Green parties.

Aren't these just random observations? Has our view not been distorted by the fact that more and more people work in information technology today? What if it is due to a set of circumstances which does not relate to the character of the IT profession itself? Statistical data might be able to give us a partial answer to these questions. In 2015 there were 80,000 IT specialists according to the statistics in the Czech Republic, 85% of whom had a university education. They count for 1.5% of economically active Czechs (6% if we focus just on those with university degree). We can, therefore, rule one thing out – the high proportion of IT professionals within environmental activities cannot be explained merely by their number in the population.

More detailed empirical inquiry encounters one specific problem. The connection between profession and nature is difficult to articulate and just few people reflect upon it. We discovered this ourselves when we turned to IT workers. Whether they were students (the interviews with those combining environmental studies and informatics were particularly valuable), university professors or they actually had job in the IT, they were all equally hesitant to make unequivocal judgements.[150]

[149] The relative financial independence combined with an interest in nature can lead to environmentally harmful behaviour - flying to exotic lands or commuting with several cars from romantic secluded places.

[150] In such a situation, honesty was particularly valuable. It is more common for uncertain respondents to start to improvize or self-stylize in their answer. We shall discuss this in Chapter III.5.

An explanation for the environmental sensitivity of (some) IT professionals, based on a sociological look at history, is offered by the media historian Fred Turner (2006). He traced the origin of the computer revolution to the same place which gave rise to the modern environmental movement – in the Californian counterculture of the 1960s. From its initial development after the Second World War, cybernetics as a scientific discipline has evolved in conjunction with ecosystem thinking, especially at the level of microbiology. For Turner, the key figure who brought this connection to the public eye was the bohemian techno-optimist Stewart Brand. Brand, a trained systems biologist, began travelling across the US in the 1960s in the Merry Pranksters' legendary 'acid' bus. He gradually became the figure who most influenced the hippy subculture's relationship with modern technology. Thanks to him, this technology, which had previously been seen as a symbol of the Cold War and state power, now acquired an entirely new association with freedom, youth revolt and self-realization.

The most visible expression of Brand's work was at the end of the 1960s with the publication of the *Whole Earth Catalog*, which, half a century later, Steve Jobs compared to the Google search engine in paper form. This hefty tome, with its front cover featuring the iconic image of the Earth as seen from a spaceship, was inspired by cybernetics, communalism and the DIY principle, and became the handbook of the back-to-the-land movement of the 1970s. A few years later, Brand was instrumental in mediating the works of systems thinkers such as Gregory Bateson, Norbert Wiener and Buckminster Fuller to the hacker subculture,[151] pioneers at the brink of a revolution in personal computers. In 1984, for example, he organized an event which was later called the 'Woodstock of the computer elite' for bringing together several hundred computer enthusiasts. Even as he has grown older, Brand has not abandoned the idea that information technology can lead to a sustainable future (2010) – at present his ideas place him in the group of ecopragmatists (see Chapter III.2.).

The hypothesis that IT professionals have a particular affection for nature is strengthened by the presence of theoretical mathematicians, computer scientists and cybernetics amongst the leading figures of the environmental movement. One such representative is Juraj Lukáč, the enthusiastic leader of the Slovak NGO Lesoochranárské zoskupenie VLK (Forest Protection Movement WOLF), a trained electronic engineer who has worked for years in cybernetics. Lukáč argues that it is crucial that theoretically educated IT professionals are trained in abstract and systems thinking. With his unmistakable zeal he explained: "Biologists have a very simple understanding of nature. When they learn the Linnaean classification to distinguish a small woodpecker from a large one, they then think they understand nature. A mathematician doesn't think that way. Mathematicians and IT professionals are able to see the complexity in the links in a natural system and understand the risks of human intervention.

Lukáč is not perturbed by the fact that in practice the most visible connection IT workers have with nature is through sports activities: "I was just the same. Just leave them to it a bit longer – even I needed ten years. But it will come to them. There is no need to convince them or teach them because they already have that mindset. All they need to do is draw back the curtains. They run about with their GPS's but they don't have the barriers in their minds that, say, foresters have. You can explain things to a forester for twenty years and they'll still not understand." Juraj Lukáč doesn't only fall back on his own experiences – he estimates that out of six thousand active supporters of VLK half of them are mathematicians and IT workers.

In addition to Lukáč, we could add other key figures from the Czech environmental movement and its intellectual milieu with a similar professional background – for example, Josef Vavroušek, Ivan M. Havel, Jan Sokol[152] and Zdeněk Neubauer. The philosopher and biologist Zdeněk Neubauer wrote engagingly on the deeper epistemological dimension of the links between cybernetics and the living world. Although his arguments are based on different starting points from those of Juraj Lukáč, they lead to similar conclusions. In his view, cybernetics, the science of "the skills of the helmsman"[153] in managing complex systems, provides a more inspiring and appropriate view of the living world than biology, which is governed by a mechanistic paradigm:[154] "The scientific and intellectual currents associated with cybernetics and informatics provide the tools and the preconditions for breaking out of the modern-day straitjacket and making humankind, consciousness and life part of a meaningful unity of the world once more. (...) The cybernetic method of asking questions (together with modern physics and generalized thermodynamics) can help rehabilitate a biological (biomorphic) view of the world" (Neubauer 2002, pp. 24, 26). After all, even some IT workers agree with the idea that at the heart of their profession lies the precondition for understanding environmental issues. For them, rationality based on a focused analysis of the problem, or work at a high level of abstraction, influence their

[152] Professor Jan Sokol – a philosopher, anthropologist and mathematician – responded cautiously to our hypothesis. In his view, carbon reality is in fundamental conflict with "silicone thinking". The world of nature has "a mind of its own" which IT workers have trouble with.

[153] Etymologically the name of the discipline comes from the Greek *kybernētēs*, a term for a director, leader or helmsman.

[154] In order to explain the differences in both these views of reality, Neubauer draws attention to the meaning of the words (2002, pp. 29-30). The original meaning of the Greek *mechanan* is "to act against the natural course of things", "to slyly contrive something"; for example, by a system of pulleys bring in the final act *deus ex machina* onto the stage. Where mechanistic ideas are based on ideal mathematically expressible models, cybernetic 'helmsman' thinking is based on experience and working with chance.

[155] Is it surprising that those we contacted were drawn more towards music than technology – in both passive and active forms?[OK] Brno Professor Jiří Jan, a trained radio electronics engineer, is an excellent organist: in fact, we came across organists and pianists quite often. The programmers themselves offered interesting explanations for their affinity to music. They compared their work to composing – they use individual motifs to attempt to create a unit which "is in harmony".

[156] The Czech edition included a short story by Dino Buzzati entitled A Pleasant Night. This is due to be published in English in a collection of Buzzati's works by the New York Review of Books.

attitudes towards non-professional issues, including attitudes towards nature. Of great importance to them is that irreplaceable feeling when a complex system works – and, on the contrary, the frustration when something 'irrational'[UK] stops it from functioning for political or bureaucratic reasons.

It is interesting that most of the IT specialists we managed to talk to did not share Stewart Brand's optimism. It is true they have hope in the possibilities that will be given to people through a decentralized internet network, and they trust in people's freedom and creativity. However, they do not belong among the technological enthusiasts[155] – they often have older push-button telephones and place greater emphasis on durability and endurance than on digital gadgets. At times they mentioned the fact that they have seriously considered the risk of collapse – for example, an energy blackout – and cogently speak about "the limited willingness to delegate something to a machine". This is perhaps because they have a notion of the fragility of the system caused by its complexity, they are not fans of 'smart' solutions; their mistrust in them is due to the risk of the collapse of computer-controlled systems or the risk of abuse by a 'Big Brother'.

Zdeněk Neubauer would certainly have understood such a cautious approach. He wrote about the paradoxical situation whereby despite more and more powerful – almost miraculously successful – computers, the initial ambitions of cyberneticists have begun to fade. The experience they have gained over the years confirms one of the central ideas of this whole book. Despite progress, the attempt to simulate natural processes or to develop artificial intelligence continuously comes up against the "astonishing limitation of computer technology" and the "unpredictable difficulties with the majority of elementary intelligence functions whiché govern 'every small child'" (Neubauer 2002, p. 40). Nevertheless, many cyberneticists of the current generation are still optimistic. They believe that this is just a small stumbling block on the road towards the "second machine age". Zdeněk Neubauer would undoubtedly wager that these cyberneticists would also have to – sooner or later – confront the peculiarity, unpredictability and probable unknowability of living systems.[156]

III.5. Faithfulness to words and ideas, the problem with comparisons

'Voluntary simplicity', subcultural codes

We are slowly approaching the fourth section with the report on how the life of the Colourful has changed over the past twenty-three years and how their children have grown up. Before then, we shall briefly examine whether the findings from our research are in line with what has been written about environmentally friendly lifestyles abroad. But why only "briefly"?

The problem is not that there is a shortage of international literature about environmentally friendly lifestyles. The difficulty lies in the fact that there is too much and that it is not easy to differentiate academic books from those by mere enthusiasts. It is the case for environmental topics in general that we have to be especially wary and distinguish whether the author is not confusing dreams with reality.

Dubious texts include those which borrow the phraseology of the research subject and its implied value system, as they show a lack of insight and scepticism in the author's research. This can be seen in the example of the phrase 'voluntary simplicity'. This had become popular amongst social, gender and environmental activists before it was absorbed without further examination into the lexicon of texts on environmentally friendly lifestyles; into texts by activists, but also – and herein lies the problem – texts by academics.

Why did the word 'simplicity' become popular in environmental ideology and in the vocabulary of environmentally friendly lifestyles to the detriment of 'modesty'?[157] This is surprising: 'modesty' is very appropriate for expressing an approach based on the low consumption of consumer goods. The word 'simplicity' does not contain such meaning and could also mean refined luxury.

The reason for the preference of the word 'simplicity' may be psychological, as 'modesty' suggests a lack of material goods, while 'simplicity' has more acceptable, even attractive connotations for people today. This fact is also important sociologically and has historical roots: The idea of simplicity is linked to the aesthetic idea of the life of the elites, which has been cultivated over millennia. Names spring to mind such as the spiritual teachers Buddha, Laozi and Confucius, the ancient Greek and Roman philosophers Aristotle, Plato, Cicero and Seneca, the famous ascetic Diogenes, and the proponents of pastoral simplicity Virgil and Horatio. Over different periods the simple life has inspired,

[157] If we used search engines for terms such as 'voluntary modesty', 'voluntary moderation' or 'voluntary frugality' – we would receive relatively few results. However, our search engine would be overwhelmed by the phrase 'voluntary simplicity'.

for example, Marie Antoinette's Hamlet, the pose of the Romantics of the 18th and 19th centuries, as well as the intellectual basis of the modernist architecture in the last century. It might be said that every period of prosperity is accompanied by the desire for a simple life.

The phrase 'voluntary simplicity' was first used in connection with the environment by Duane Elgin in the 1980s. He had borrowed the phrase from Richard Gregg, a Quaker and student of Gandhi. The emphasis on simplicity has found its way into the titles of dozens of books whose authors either sympathised with, or were members of, environmentally passionate communities. We could name numerous influential books such as Bill Devall's *Simple in Means, Rich in Ends: Practicing Deep Ecology* (1988). Voluntary simplicity became an ideological expression for several alternative groups who hoped to turn society away from the 'industrial era' to the 'ecological era'. It implicitly contains a spontaneous conviction concerning an internal locus of control.[PK]

Thanks to its broadly shared and emotional connotations, voluntary simplicity reliably fulfils the function of a slogan or cultural code, or in our case, more of a subcultural code. It allows for the smooth communication within a subculture of environmentally engaged individuals as well as communication with the outsiders. However, voluntary simplicity largely fails to characterize the actual nature of environmentally friendly lifestyles. This predisposition reduces the scientific value of many texts on environmentally friendly lifestyles and the possibility of comparisons with our research, which is what we are seeking.

Moreover, today's main theoretician of voluntary simplicity, Samuel Alexander, contradicts himself in the introduction to the anthology *Simple Living in History* when he writes: "Anyone who has attempted to actually live simply, of course, will know very well that doing so is not very 'simple' at all" (Alexander and McLeod 2014, p. xiv). This is because the normal connotation of 'simple' is understood as 'easy'. This equation, though, does not apply in an environmental context.

In Chapter III. of the book *The Half-Hearted and the Hesitant* I presented nine dimensions which demonstrate why the ideology of voluntary simplicity does not describe the real characteristics of an environmentally friendly lifestyle. Here I will give a brief example from everyday life:

The structure and operation of society today is complicated, but the life of the consumer is simple. Entire branches of industry are dedicated to simplifying their existence. Products which in the past would have been made through difficult processes at home can now be easily purchased. Shops are filled with ready-made and processed food, while people in the countryside no longer have to be bothered by the complex production of their own food. Self-sufficiency, which is a characteristic of environmental friendliness, is in sharp decline. In short, an environmentally friendly lifestyle is not simple. What's more, everyday life is not even simplified by the gentle instructions of the lists of 'tips to help the planet' – indeed, they often complicate matters. There are a host of other similarly positive sounding, but at the same time, scientifically dubious codes, including 'downshifting'[158] and 'permaculture'. With little in the way of reflection they are then inherited by generations of alternative lifestylers, which demonstrates their faithfulness to ideas.[UK]

Lukáš Kala discovered some charming codes or identitary words,[159] when as part of his thesis on 'green singles' he examined the profiles of clients on internet dating sites. As we will see in the TEXTBOX at the end of the chapter, during their searches, those clients who were environmentally motivated tended to look for partners using stereotypical words and ideas rather than references to actual behaviour or specific characteristics (Kala 2014).

The fact that members of the same group can easily orientate themselves using subcultural codes, that they follow their 'scent' and quickly and reliably find one another, was also used by researchers at the Melbourne Simplicity Institute (sic!) when they were looking for a methodology to find respondents for global quantitative research into environmentally friendly lifestyles (Alexander and Ussher 2012).

A few words on community faithfulness

Due to the difficulties in interpreting the lifestyles of the Colourful, it occurred to us that life in communities or communes could be a suitable referential framework.[160] We thought that they could represent a kind of model because their environmental friendliness has a distinct and relatively stable form. In addition, from a methodological perspective it is important that academic texts on life in communities are usually based on participant observation rather than interview techniques which are subject to self-stylization (Kirby 2003; Ergas 2010; Blažek 2015 in the Czech Republic).

With some simplification it is possible to distinguish two main types of community – so-called *cohousing,* which is characterized by shared living and shared lifestyles, and the *ecovillage,* which strives towards

[158] Downshifting is an effort to spend less time at work and more with friends and family and on personal development (Levy 2005).

[159] Alena Podhorná-Polická examined the 'identitary neologisms' used by the youth in French suburbs. She defined them as words or phrases linked with a high degree of expressiveness and frequency, which are totemized by a specific group of speakers (2011).

[160] Jan Blažek (2015) describes a commune as a community in which the members not only share expenses but also income.

[161] Despite the help of numerous volunteers, agricultural work is hard for communities, even for subjective reasons which are less obvious: For example, the need to look after the running of the farm on a daily basis, in particular of the domestic animals, is at odds with the wanderlust of community members with their migratory tendencies, love of travel and improvisational 'communitarian nomadism' (Maffesoli 1997).[MK]

[162] Our research into the Colourful also confirmed this as an indisputable fact.

[163] We will come to them in Chapter IV.3.

nature-friendly farming. The environmental benefits from both of these forms, or if we like, their faithfulness to nature,[UK] lie in a specific lifestyle which is modest in terms of the consumption of material goods. It is a rare example of *intentional modesty*. Thanks to the farming in ecovillages, the link to nature is clear, close and direct. If life in ecovillages can achieve self-sufficiency in food, it would represent a radical variation on environmental friendliness – it is actually its concentrated version which puts great demands on everyday life practices and on the strength of your attitudes. To add to this: It follows from the nature of farming activities that farming communities are similar to one another, which is why ecovillages could theoretically be a yardstick for other forms of environmentally friendly lifestyles.

In reality, however, the constant form and concentrated environmental friendliness of communities often fade away. Ecovillages which aim for self-sufficiency are placed under severe pressure.[161] Given the general growing unpredictability of natural processes and today's economic and agronomic conditions, long-term nature-friendly farming is practically impossible.[162] The common fate of rural self-sufficient communities is that over time their activities decline and are replaced by a wide spectrum of other activities.[PK] Community members who wanted to be farmers end up as social workers, they help local farmers distribute food to the cities, they take part in local conservation activities and try to introduce translation services and educational courses into already saturated markets (Forrest and Wiek 2015; Gálová 2013).

Moreover, Jan Blažek (2015), who has spent a long time studying community life in Slovakia, Denmark and Portugal, suggests that any idea about communities having similar lifestyles is wrong. In his view, the communities are as diverse as the structure and make-up of their members, and as diverse as the external conditions they live in. He has even found some old-new trends which differentiate the range of community activities – for example, looking after community gardens, projects for the development of abandoned and semi-abandoned rural settlements; in Portugal a significant back-to-the-land trend can be observed. Therefore, not even Blažek's field observations can confirm our idea of the ecovillage as a homogenous and stable reference point for an environmentally friendly lifestyle.

But the communities certainly do offer us something important – the opportunity to see the ideological bases of radical attempts to change lifestyles which could provide a framework for our comparisons with the attitudes and motivation of the Colourful.[163] The language used to express these communities' attitudes was formulated in the 1960s and 1970s[164] and has been reproduced and disseminated with a faithfulness reminiscent of the unquestioning approach of translators and publishers to Biblical texts. Their foundation charters, community programmes, statutes and manuals are strikingly similar, despite being passed off as the original product of lively group discussions. They have an emotional urgency and intellectual radicalness which the individual environmental activist would be hesitant to display.[165] They are formed in a climate of mutual support and relative isolation from society, displaying an enthusiasm which hasn't yet been eroded by life experiences.

The documents show that although the motivation behind communities in the Czech Republic and Slovakia has some teleological elements,[MK] community members do not aspire to the political role to change society; the 'internal locus of control'[PK] has only been weakly developed there. Two extracts from their "about us" pages show a desire for a community idyll rather than involvement in a political movement:

"We are a group of like-minded people who in the future would like to live in the style of cohousing or ecovillage. We are united by our positive relationship towards nature and everything natural (but this does not mean that we have turned away from technological innovations). We would like to bring more creativity and shared experiences into our lives. We would like our children to be able to live life using all their senses and experience it from more perspectives than is usual today. We would like to revive old traditions and mark the passing of the year through shared celebrations. And finally, there is also the opportunity to earn a living by doing something we all find fulfilling and enjoyable, and, best of all, in the very place where we live" (Kmen 2013).

We can read similar rhetoric, this time with an environmental emphasis,[166] in the presentation for another project, Pôdne Spoločenstvo Olšinka, which has been operating near Bratislava since 2010: "We are united by a sense of responsibility for the impact of our behaviour on nature. We perceive and respect the value of all living and non-living things on Earth, regardless of whether this value is useful to us. We strive to reconcile the provision of our own basic needs with caring for the Earth. We are gradually realizing our idea of a more environmentally responsible life. We are creating an environment and conditions which automatically and non-violently support environmental ideas and reduce demands on energy and external sources" (PSO 2010).

[164] Naturally, the historical roots of attempts to create modestly living communities go back much further and were mainly religious in character. They include the Quakers, the Amish and in particular the medieval monastic communities. In the 1920s and 1930s, there existed communities of a secular nature (Pausewang 1980).

[165] Let alone the Colourful type of person.

[166] A typical inclination towards the principles of permaculture is present there (Mollison 1988).

Faithfulness to ideas as a methodological issue.
The limitations of comparisons with the research
of the Colourful

Towards the end of 1992 I was processing newly filmed interviews on voluntary modesty when it struck me again how difficult it would be to interpret them. I began to wonder if it might have been better to have used a standardized questionnaire which would have prevented the in-depth interview themes from splintering into dozens or even hundreds of details which would be difficult to generalize. It was hard to find a common denominator in the recordings which would allow for a routine sociological analysis, including comparisons with other research work on a similar theme. Some of the shared features in the everyday behaviour of the respondents were clear (for example, their relationship towards shopping, aversion to travel and the high level of involvement in the community's civic life), otherwise the group was characterized by a large level of heterogeneity. I opted for a cunning solution, which was almost unacceptable from a methodological perspective: I simply called the respondents *the Colourful*.[167]

On the other hand, I greatly appreciated the irreplaceable aspect of in-depth interviews as they contained nuanced information on everyday life, which could be expressed thanks to the atmosphere surrounding the interviews and the trust between the interviewer and the respondents. In Czechoslovakia in 1992 we were inundated with neoliberal propaganda before we were ready to counter it with knowledge of environmental ideology from the West. The nascent Czech forms of environmentally friendly lifestyles had not yet been able to create an ideological vocabulary. As a result, our discussions on nature-related issues were spared 'green clichés'. They represented that rare opportunity when communication independently stumbled upon apt words.

Researchers had been studying environmentally friendly lifestyles in the West under different conditions, where an ideological background of alternative lifestyles had been developing unimpeded since the 1960s. It might have escaped the researcher's notice that their questioning probably elicited images which had been contaminated by a long-held ideology. The research findings might have been influenced by self-stylization,[PK] which distorted the results about attitudes and lowered the validity of statements concerning everyday behaviour. This research pitfall in such a 'schooled' field lurks for the researcher even when creating a sample.

[167] The next two phases of the research in 2002 and 2015 revealed that the heterogeneity of our respondents' lifestyles had increased even more over time.

A broad picture of environmentally friendly lifestyles is given by Mary Grigsby in the book *Buying Time and Getting By* (2004).[168] At first sight this publication appears to be a suitable comparative framework for our research into the Colourful. However, when we look at her method for collecting data and selecting her respondents, we find that a reliable comparison is not possible: The author chose her respondents from several workshops focusing on 'simplicity,'[169] and used the methodology of participant observation. She then carried out in-depth interviews with the most active participants. She was interested in the literary sources of their attitudes and it is little wonder that she found some enthusiastic classics relating to voluntary simplicity (pp. 27–29).

I consider the most reliable source of information to be sociological studies where the features of an environmentally friendly lifestyle are of secondary importance to the study's main objective. For example, there have been several studies on gardening (Bhatti *et al.* 2009) and the use of domestic technology (Lorenzen 2012). There was a recent noteworthy study of American families who live without television sets (Krcmar 2009).[170]

On the other hand, I consider statistical surveys on attitudes towards the environment to be unreliable. Respondents gush about their love for nature when in reality it transpires that most of them are not interested in nature protection. A reliable, indirect picture of environmental awareness and responsibility in the Czech Republic appeared in the autumn of 2015 in connection with a scandal surrounding illegal software which affected the volume of fumes emitted from diesel engines. For those of you who don't remember the affair:

The Volkswagen car company had been forced by European legislation to equip diesel engines in cars with a device to reduce NO_x emissions. The downside to altering the engines was reduced performance and increased running costs. Volkswagen, therefore, decided to fraudulently bypass emission measurement by using software that was able to distinguish the test mode and adjust the engine to meet the standards during the test. When the car was running normally it emitted up to forty times the amount of NO_x permitted. We don't know how the owners of eleven million modified cars across the world reacted, but according to a news report on Czech Radio and an interview with the Minister of Transport on the Radiožurnál station from 24 September 2015, Czech drivers – predominantly owners of Škoda vehicles, which is part of the Volkswagen group – did not intend to

[168] The method by which Grigsby created her research sample corresponds with the book's subtitle *The Voluntary Simplicity Movement*. We will discuss the label 'movement' and its relationship to the life of the Colourful at the end of Chapter IV.3.

[169] She writes about "simplicity circles".

[170] There have been many master's and PhD theses at the Department of Environmental Studies at Masaryk University on the environmental aspects of certain lifestyles, for example, eating in luxury restaurants (Geryková 2010), shopping in second-hand stores (Lukašíková 2009), purchasing cut flowers (Kosová 2006), mushroom picking (Langová 2012) and the life of the Catholic Order of Saint Clare (Popelková 2013).

visit a garage to ask for their engine to be modified. They were not bothered if their car was contributing to air pollution; what was more important was not to reduce the performance or increase the cost of driving.[MK] And it was also an inconvenience; the trip to the garage to fix the problem was described by the minister as a 'trek' which people would want to avoid (ČRo Radiožurnál 2015). I wonder how those people responded when asked in public-opinion polls whether they were concerned about 'a healthy environment'?

My conviction regarding the suitability of qualitative approaches does not prevent me from being interested in the results from sociological research based on statistical methodology. So I looked to see if I could find any quantitatively based inquiries into environmentally friendly lifestyles to compare with the research into the Colourful. In 2012, Samuel Alexander and Simon Ussher conducted a rare quantitative international study entitled "The Voluntary Simplicity Movement", which was published in the *Journal of Consumer Culture* (Alexander and Ussher 2012) and which is often cited. The sample consisted of 2,268 respondents: 970 from North America, 871 from Australia, 147 from Great Britain, 108 from other countries in West Europe, 77 from New Zealand, 4 from Japan and 97 from other parts of the world. They answered 50 questions which focused on the respondents' demographic background, their consumption, income, values and motivation, their relationship towards politics and the community, and finally their feeling of happiness.

Is it possible to compare the results from this global survey with the findings from the study of the Colourful? I soon realized that even on this occasion a comparison was only possible in a limited way. The reason for the low-level of comparability did not only lie in the principal difference between quantitative and qualitative research, but also in other differences in the theoretical, methodological and technological level of the research design.

The two studies differed from the outset in the way in which the criteria for environmentally friendly lifestyles had been established. In the research by Alexander and Ussher, voluntary simplicity was repeatedly defined as a lifestyle of "reduced or restrained income, consumption, and/or working hours" (pp. 71–72), as though these phenomena were necessarily linked. The basic criterion for the selection of our Colourful was lower consumption than the household could afford; income did not play a decisive role. Although the prototype of a voluntarily modest household minimalizes both input (income) and output (consumption), in our research we also recognized households with a high income and low consumption.[171] A low overall *household metabolism was key for the creation of the Colourful sample.*

Samuel Alexander and Simon Ussher's research significantly differed in the way the sample was chosen; this was done by self-selection. The researchers contacted organizations, websites and blogs which were linked to voluntary simplicity. They also turned to academics, environmental-education teachers and activists who had declared their link to the simplicity movement. The respondents were contacted through social network groups with themes such as 'voluntary simplicity', 'simple living' and 'downshifting'. The authors of the research assumed that most people who live an environmentally friendly lifestyle and are interested in the topic would occasionally use these online resources.

For our study of the Colourful, we approached the sampling in an entirely different way. We created a list of possible respondents, which was mostly based on references given to us by students from my seminar on environmental aspects of lifestyles. Their suggestions stemmed from our definition of a voluntarily modest lifestyle and structural demographic characteristics.[172] They were also given instructions not to include environmental activists amongst their list of respondents. Our respondents would never have been part of Alexander and Ussher's research because they did not identify with any ideologies, let alone an environmental one, and they were not associated with environmentally themed social networks. Similarly, those respondents from the Australian research who declared environmental interests would not have been included in our group of the Colourful.[173]

Another important methodological difference was the range and depth of the questioning: The method we used was an interview lasting as long as two hours; the Australian research consisted of an online questionnaire. When interpreting the significant differences in the results of the two studies – we will look at them in more detail in Chapter IV.3. – we have to keep in mind that they may have been significantly affected by the methodologies selected. This also applies to researching motivation, a phenomenon which this book pays particular attention to.

[171] It was very unlikely to find such households in 1992. There was one case of a family of a famous Prague artist. In the research in 2002 the probability of their occurrence increased, and even more so in 2015.

[172] We attempted to keep roughly within the quotas from the demographic structure of the population of the Czech Republic.

[173] They would have been classified in the Green group.

Subcultural codes in an international 'green dating site'
(from the notes of PhD student Lukáš Kala)

MEN:

Green arms wide open: "I love God, nature, and all living beings and I am seeking someone who feels the same." *lovemystic76*

Save Mother Earth through a compassionate diet: "I help the planet by following a compassion-based diet and lifestyle. We help the planet by simply giving her our love, and we in turn are blessed by our Earthly Mother for our compassion." *lovemystic76*

Culture is killing the planet, the loneliness of a community farmer: "I think this culture is killing the planet, in fact I don't think it, I know it, but somehow I have retained the ability to joyously laugh and love. I live on 220 acres in the mountains of West Virginia with 12 of my dear friends... I have been a bit frustrated lately with trying to find love." *Peiro*

I will rescue you from the iron cage of modernity: "I am an Anarcho-Communist (message me for more explanation on my political views and why). My activism roots come from animal rights and will always be one of my strongest passions. I am also an enviro activist and believe in the liberation of all land and oppressed beings, no matter the cost." *Anarcho-Communist*

I'll bite you if you try to talk to me about meat: "I'm also a firm opponent to speciesism. I'm tired of being surrounded by people who consciously ignore or pick apart my high ethical standards. I won't settle to date any more meat-eaters!" *LDN*

WOMEN:

Seeking someone to build castles in the air: "I'm a hard-working lady with a fire to make a positive impact on my planet and my community. I strive to be more sustainable in my life. My dream is to find a man who shares my vision for a self-sustaining homestead." *DesertGreenGirl*

Everything is connected: "My belief is very holistic. I feel that everything is energy. I highly respect all old nature-based spiritual beliefs and shaman practices. I believe there is spirit in everything and everything is connected." *lenae*

No more shamans: "I value local, green, sustainable living but I don't vibe with what often comes with it—the crunchy gaia/shaman/drug use paradigm. In short, I'm searching for someone who isn't a sheep, someone who doesn't need others approval, someone who sticks up for his beliefs and has a strong backbone, a humanitarian with a strong sense of morality." *Racheljune3*

No-one understands her. She only wants to live responsibly: "Trying to live in the most responsible way appears difficult almost when you're not supported by your family or friends. That's why it seems important for me to find a person who will share my views without saying I'm a fundamentalist or such things. I'm just trying to protect the space I'm living in and my own health in the most simple and natural way." *LEROCHKA*

She has gone down the rabbit hole in search of knowledge about life and has decided not to return! "I'm keen on conscious living, deep, heart-centered communication, permaculture, sustainable communities, innovating regenerative business models and challenging the consensus reality (including my own—without proselytizing). My curiosity for knowledge, spirit, healing and wisdom has led down many a rabbit hole and will continue to do so." *justintilson*

IV

The Colourful over the course of time

IV.1. The initial report: the Colourful in 1992

(Reprinted from the book *The Colourful and the Green: Chapters on Voluntary Simplicity* [1994]. H. Librová has made only minor changes, adding some indices referring to the theoretical keys.)

One sociological study

In June 1992 the United Nations Conference on Environment and Development was held in Rio de Janeiro. The delegate sent by Czechoslovakia was Josef Vavroušek, the chair of the Czechoslovak government's Federal Committee for the Environment. Josef returned clearly affected by the attention from Western representatives, who saw hope in the post-communist countries. They believed a new chapter was opening up for Czech and Slovak societies, offering the chance to avoid the same mistakes made by capitalist societies, which had led to consumerist and environmental disaster. At a time when expectations for rising living standards were high, Minister Josef Vavroušek came to me with an unusual request – he would welcome it if sociologists could answer the question: Is there the potential in our country, or is there a model, to promote an environmentally friendly lifestyle? If so, who are the people who voluntarily content themselves with low material consumption and what kind of lives do they lead? It is worth noting that this inspirational conference took place in Rio from 3 to 14 June. In September of the same year (!) I set out with the filmmakers Ivan Stříteský and Aleš Záboj on our first field trip.

An environmentally friendly lifestyle is a wide-ranging term. For the purposes of empirical research we had to narrow down its meaning. We have no doubt that people are behaving in an environmentally friendly way if they are actively involved in nature conservation – they plant trees and protect migrating amphibians – or if they abide by the principles of sustainable household management – they clean their clothes using phosphate-free washing powder and take used glass to be recycled. However, in our research the main criterion was *people's efforts or willingness to live modestly,* regardless of whether they belonged to an environmental movement. Nature conservationists were also included in the research sample if they fulfilled this main criterion. They were placed in a special subcategory, the Greens, principally for comparison with the Colourful.

Our criterion can also be described as people's willingness to live under conditions which are unfavourable materially.[MK] The reader might imagine that this mainly applies to old people. However, for the majority of them this is not a deliberate choice but has come about through inertia or the habit of thriftiness, or out of necessity. Therefore, we have not included this type of modest person in our sample.[174] Similarly, our research was not concerned with people who had been forced by circumstances to reduce their consumption; in the majority of cases, we would expect them to seize any future opportunity to consume more. Our study focused primarily on cases where the lifestyle was characterized by *voluntary and intentional modesty*.[175] In a society-wide context this represents a real change in values, which for most people are directed towards greater consumption.

In our project we did not consider cases which a preliminary examination revealed to be on or beyond the border of psychological norms. The research also excluded people whose lifestyles were governed by the rules or ideology of a group, *e.g.* a sect. For this reason, the research did not take into account the otherwise highly relevant issue of environmentally friendly lifestyles in communities.

When constructing our research sample, we focused on people who had already started a family. We can assume that their lifestyle is more or less stable and is not merely a momentary whim. In the case of single people, we looked for individuals who were over thirty years of age.

From the start it was clear that the main problem for the research would be the methodology for finding its respondents. Naturally, there were no lists of people matching our criteria, so we opted for the method known as 'snowball sampling'. We used data from 'informers'– sociology students, mayors, priests, environmental activists *etc.* – who had expert knowledge of the social situation in their neighbourhoods. We attempted to eliminate the danger of sampling bias[176] by selecting our relatively large number (23) of informers from different social and geographical backgrounds.

As we had no reason to believe there would be significant differences in environmentally friendly lifestyles in different areas of the Czech Republic, and as any attempt at geographical representativeness would have been overly complicated (in terms of funding, time and organization), we chose six areas for our field research: West Bohemia, South Bohemia, Prague, the Bohemian-Moravian Highlands, Brno and its surroundings, and East Moravia.

A contact list was drawn up with 113 addresses, of which 47 – those belonging to the relevant areas – were selected. In view of the fact that

[174] Nevertheless, the modesty of old people was a source of inspiration for our respondents, as we shall see.

[175] 'Intentional,' 'voluntary' and 'conscious' simplicity are terms used quite freely by sociologists and journalists in the West. The authors fail to take into consideration the different shades of meanings these adjectives convey. In our case too it only became apparent in the course of our research why it was necessary to distinguish between 'intentional' and 'voluntary' modesty.

[176] Even so, we cannot rule out the possibility of some shifts in representativeness due to the sampling method. For example, it is possible that 'intentionally modest' people who are not very communicative did not make it into our sample. They might have escaped the attention of the informers or struck them as unsuitable for research interviews. Our research can therefore be seen as more of a typological study.

the number of interview refusals was surprisingly low (three in total, two of them being private farmers who gave convincing reasons for their unwillingness to be interviewed, explaining that they were too busy with the autumn season), we can rule out any bias arising from self-selection. A total of 44 in-depth interviews were carried out with 70 people (in some interviews there was more than one person taking part).

The research subject was nearly always a family household or in some cases an individual. In the interview, the family household was usually represented by a married couple, sometimes accompanied by their adult children. This set-up proved to be valuable in methodological terms as the communication between family members revealed important facts concerning their lifestyle, motivations, social roles and so on.

From the outset, the objectives of the project and the particular features of the subject area and the sample precluded the use of usual methods of standardized questioning. The basic methodology for data collection was an in-depth interview which was carried out according to a pre-tested interview guide.

It should be noted that the scepticism concerning research based on ascertaining attitudes, which we otherwise share, can be weakened in the case of our research. This is because our respondents were better equipped to resist the well-known temptations which usually lie in wait: self-stylization, opportunism in relation to the interviewer and the desire to identify with generally shared values. The respondents had firmly rooted attitudes towards life. It is important to emphasize that these are people with a relatively high degree of resistance to the influence of public opinion and other social pressures. Most importantly, due to our actual presence in the households, the content of our interviews was confirmed by direct observation.

At the end of this short chapter on the research methodology, I would like to thank my sociology students, some of whom helped me technologically process the information from the interviews. I also drew inspiration from some of the students' seminar papers for my research hypotheses, as I have shown in the bibliography.[177] I acquired several contacts for the research sample thanks to the ingenuity and helpfulness of my younger colleagues. My work benefitted from all the discussions which helped to keep me on the right track.

Who are the people willing to limit their consumption?

In spite of all the research doubts, and in spite of all the reservations which could be expressed about some of the details of the methodology used, we can now confidently state: There are people living in our society who reject the notion of a general trend towards high consumption. There are families and individuals in the Czech Republic who voluntarily or intentionally lead modest lives.

These conclusions immediately raise the question of how common these lifestyles are in relation to the whole population. However, the research design prevents us answering that question; indeed, it would be very difficult to find a method of counting the cases of intentionally modest lifestyles. From our experience, though, of looking for respondents we would say that they are not a complete rarity in this country.

[177] I have cited the works collectively (FF MU 1992). In particular, I have cited works by B. Rozbořil (1992) and M. Misíková (1994).

The *age structure* of our sample ranged from 22 to 63 years old. We believe that the typical age for the Alternatives[178] matching our criteria, who can also be assumed to have mature, long-held attitudes, is between 25 and 40 years old.

The relatively young age of our respondents begs the usual question from the sceptic: How long will their idealism last? This is not without good reason. It can be a sobering experience when a young person with romantic ideals has to confront the realities of life. Do these people go on to have the same living standard and lifestyle as the majority? Our research has not been constructed to give a considered answer to this question. However, it does suggest that the character of the Alternatives does not fit with the transient and bitter experience of many communities and movements which is described in sociology books. It indicates that the chosen lifestyle has greater permanence. The reasons may lie in the fact that it is rooted in family life and in a transcendental realm, independent of role models and the fashions of the day, and in its flexibility – qualities which we shall discuss later.

There were two distinct Alternative cases in our sample which had endured for more than 12 years. In both of these families it was evident that alternative activities had long been practised and improved upon.

One interesting point to come out of our findings was *family size.* Our sample contained a large number of families with above-average numbers of children, or very young families with one child who were planning to have more children. It would appear that families with five or six children were not exceptional here. This undoubtedly relates to the large number of Christian couples in the sample, but also to the nature of this way of life, which is orientated towards family and social life.

[178] In the 1994 book I inaccurately described the Colourful as 'Alternatives'.

131

On the other hand, our sample also included an interesting category of middle-aged single men. They were over-represented in terms of statistical probability in the Greens subgroup. It is also worth noting that these were not urban cases. We wondered if nature conservation might be a sort of personal refuge for a certain type of man, an escape from the pressures of a society where the social prestige of a bachelor is low.[179]

As expected, environmentally friendly lifestyles are often enacted within *multigenerational cohabitation* – another feature that goes against the generally accepted modern notion of the family. The efforts to cohabit harmoniously with older parents was characteristic of some of our respondents.

The *education and professions* of those who advocated an environmentally friendly lifestyle proved to be a very interesting issue. The vast majority of respondents had a secondary or tertiary education that was not used in their professional life. Typical career paths for men included an artist who was a trained civil engineer, a university-educated zootechnician who was looking after ungulates in a zoo, an economist selling religious literature, an agronomist working at a sawmill, a graduate from a mining university who became a woodcutter and is now a deputy mayor, a university-educated maths teacher working to protect landscape parks, and a graduate from an agricultural college and a professional zoologist who were now private farmers.

The majority of the women had a secondary education and a surprisingly large number had a secondary medical education (nursing) or a qualification in special education. Was this a coincidence? Almost without exception, the women from our sample were not making use of their qualifications and were at home with the children, which is easy to understand given the number of children.

In addition to these cases where people were not making use of their university qualifications, there were also many instances of people leaving university after first or second year, in particular those specializing in science or technology. This was most commonly the result of disillusionment with the course content or the standard of teaching. Students looking to strengthen their relationship with nature who attended agricultural and forestry colleges were surprised by the number of technical and science subjects (maths, physics, programming) and often transferred – at first sight illogically – to a humanities course or left higher education altogether.

[179] The single man (or woman?) in nature conservation could be a subject for an original socio-psychological study.

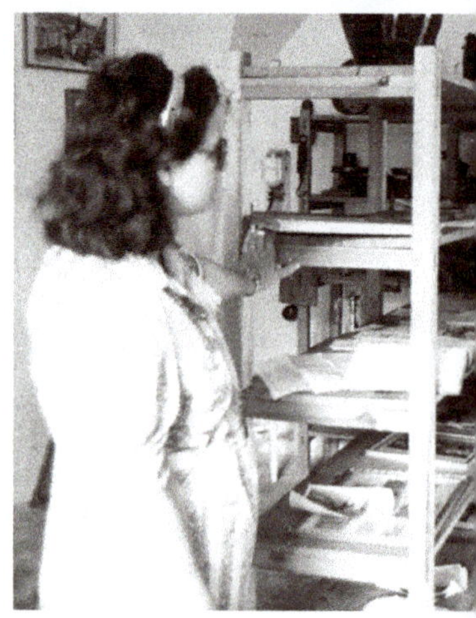

[180] Interestingly, this trend is in keeping with Ernst F. Schumacher's ideas on the rediscovery of manual work (1973). It is even more reminiscent of the Romanticism of John Ruskin and William Morris and the pre-Raphaelites' views about the renewal of society through crafts and applied art.

[181] However, unlike people who have no options from an economic perspective, our respondents did have choices and consciously decided against using them.

[182] In *The Colourful and the Green* I only gave the respondents' initials. Permission to use their full names was only granted later on.

← The research sample included several amateur and three professional artists (Jan Svoboda).

↓ Husband and wife Lia Ryšavá and Tomáš Ryšavý set up a museum display to bring together material evidence of life in Miroslav in bygone days.

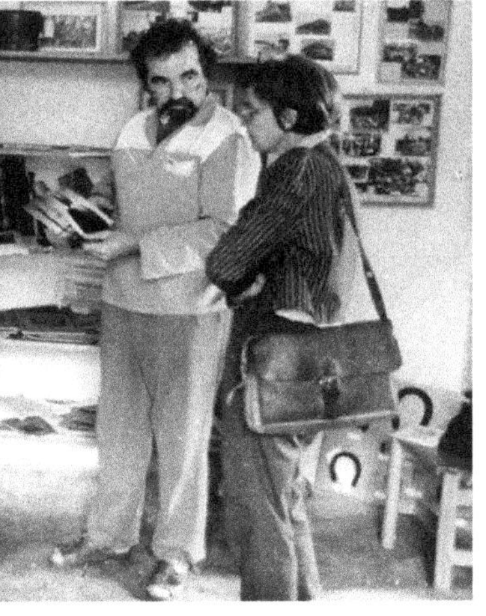

On several occasions we met with remarkable people whose highest level of education was vocational training. Nevertheless, the educational outlook of these respondents was in many respects broader than that of the average university graduate.

We might also add that the Colourful's work often revolved around craftwork.[180] Our respondents' interest in art was highlighted by the fact that three professional and several amateur artists were to be found in our research sample.

One of the most important characteristics of our modest respondents was their *residence*. Although our sample contained some people who were based in large cities, including Prague, living in a small town or village was more common. Alternative city dwellers often travelled to their weekend-houses in the country and were intending to move there permanently in the near future. There were numerous examples in our sample of people who had moved from the city to the countryside, often to what had been a holiday property or to a family home acquired through post-1989 restitution process.

Modesty in practice: bicycles, potatoes, and a dislike of travel

Most of us associate the idea of voluntary modesty with savings: A modest person saves because their income is greater than their expenditure. However, this is not the case for the majority of our modestly living respondents. Most of them don't have savings (or in many cases even a bank account), simply because they have *low incomes*. If they wanted to consume more, they would have to change jobs. They would have to adapt their interests accordingly; they would have to change their attitude towards a life which they had consciously decided on and which was of fundamental importance to them.[181] For example, it was characteristic that in the families with children, only one parent was working. These families with several children obviously did not share the opinion that 'you can't live on one salary'.

Therefore, the modesty of the majority of our respondents began with the decision to be content with a low income. "We can live with what we have. We have what we need – you can adapt to what you have," said V.R.,[182] whose family (including two teenage sons) survives on one salary and one disability allowance. When looking for a job, the financial aspect was definitely not the number one criterion. R.K., a train dispatcher in the South Bohemian borderland and an organizer of cultural events, remains faithful to his less-than-lucrative work on the railways, even though he hears stories every day about how his neighbours are getting rich working abroad: "You can't help wondering if you're a fool not to jump at the chance of working in Germany... I enjoy my work... But sometimes I regret it slightly when I see people heading over there who can't even speak a word of German." Moreover, the activities our Alternatives are involved in outside of work usually involve no financial reward whatsoever. And as we shall see, most of them are very passionate about these activities. We even met people who live with their families on the breadline but are willing to contribute to public life and pay for events organised for the community (publishing magazines, making long-distance telephone calls for organizational activities, *etc.*).

Within the context of the other interviews, this statement by 26-year-old V.F. did not strike me as unusual: "When I left the forests, I got severance pay. I divided this into three: I gave one third to my parents, another third to the Czech Union for Nature Conservation and I kept a third for myself. I donated some of that towards a Gamma Knife. I don't really need much just now... It'll be different when I marry and have kids."

From an economic perspective, our sample was divided into two unequal subgroups: The first subgroup consists of people with no financial reserves – 'we live from paycheck to paycheck'. The smaller second group has financial reserves but doesn't use them, simply because buying things doesn't interest them and in some cases actively annoys them. Like the first group, its priorities lie elsewhere.

However, it is not only about financial reserves. The economic situation can also be determined by *other reserves:* Our respondents have other reserves in the form of ownership of a small property, a house or a piece of land. Another reserve is the knowledge that parents can help in the event of a real emergency. This can be a strong psychological factor underpinning the willingness to live modestly.

From an environmental perspective, it was important that the people we visited consumed little. Actually, their key criterion was not the size of their property or how much money they had, but the *dynamics of buying and consuming.*

A special case consisted of some people from our Alternative group who put most of their resources into *renovating property.* This certainly involves a high level of consumption – building work requires construction materials, transportation, *etc.* However, this was not a case of consumption for consumption's sake or according to the principle of 'buy it and bin it'. These activities (usually renovating old rural buildings) result in objects of lasting value – not only from an architectural perspective, but also in terms of culture, history and shaping the landscape. Naturally, the respondents who are renovating properties have financial difficulties: "It's not easy with money – anything we save gets swallowed up by the house. When we wanted to buy a telephone, we had to sell the old barouche to get it installed." Many of them were heavily in debt.

Of course, money matters are obviously not the most meaningful criterion for assessing lifestyles. We can learn more from a first-hand *view of households and their furnishings.*

The limited amount of furniture was striking, and there were two visits where we sat cross-legged on the floor with our hosts. In spite of this, though, we did see a number

When something is troubling Jasmína Vaňková, she hangs out the washing – but preferably with her back to the façade of the manor house, as that only adds to her worries.

of smaller items in most of the flats which did not have a utilitarian function. They were evidently lying around because the residents had some kind of odd, perhaps even 'foolish' link to them.

Most of the furnishings had either been inherited or acquired second-hand (at the very least). In many cases we found ourselves in an environment which was clearly aesthetic and comfortable. It was also pleasing to see that our Alternatives were not anxious about adhering to the strict rituals of tidiness and order.[183] Only in a few cases was there a kind of haphazardness in the furnishing of the flat and the attitude of 'what does it matter?'.

It is interesting to compare this with the information on furnishings in Czech households presented by the Census of the Population, Houses and Flats. Generally speaking, when it came to durable goods the households we studied were significantly below average. They have the basics: a refrigerator, vacuum cleaner and washing machine (although we did come across one young woman who was happy to use a washboard). Most of the households did not have more complicated gadgets to simplify work in the kitchen - *e.g.* a blender or toaster. Almost none of them had a deep-fat fryer or a microwave oven, and they also lacked modern colour televisions and video recorders – the kind of equipment sought after by young Czech households.

On the other hand, there were some items which were commonplace in our families – a typewriter,[184] almost certainly a sewing machine, definitely a bicycle and the basic tools for a home workshop. Musical instruments, tape recorders and record players were also often found in the households.

Another characteristic was a general dislike of complex gadgets. The artist/woodcarver J.S. didn't even have a lathe for his work: "I get lost if there's a machine involved. I'm selling my skills – or lack of them." Although our respondents have relatively large gardens, most of them work on them manually. When listing his possessions, O.S. added quite gently: "The wheelbarrow is important for us." With his ideas about handicraft, E.F. Schumacher would have been proud of these Czech respondents.

However, just to be clear, many of the respondents happily used computers in their work and most of them owned a telephone (only in a few cases was this vital due to the remoteness of their home). They are not suspicious of technology in principle; they only weigh up whether it's really needed in life.[MK]

What attitudes do the people from our sample have towards cars? It's probably an even split. For those living in more remote areas, not having a car can often make life more difficult. Nevertheless, they also answered: "We don't need a car, we've got a bike," and on one occasion, with a gleam in their eyes: "We have a horse."

The bicycle is perhaps the most characteristic possession. In fact, all of the respondents said there was a bicycle in the household. In most families there were as many bikes as there were family members, which testifies to how they were used – for trips and other family events.

If objects and equipment break down, they are repaired time and time again, either by the owners themselves or at local repair shops. There were also many cases where the owners didn't have the time or inclination to fix things or have them repaired. Then the objects would lie as they were in a cupboard or corner of the house. However, this was not seen as a reason to buy more up-to-date equipment; instead it was evidence of what you could do without.

We also met some people who were thinking of getting rid of household items and not replacing them. They weren't exactly novices in the field. At some point they had cleared out their curtains, their car, the television, and now it was the turn of 'useless' items of furniture: "That one's days are numbered," said the Euro-Indian V.F. gloomily, pointing in the direction of the sofa.[185] However, it is necessary to add that this radicalism did not extend to 'Indian' objects: animal skins on the apartment walls, totems, head-dresses and other ceremonial objects.

In the eyes of many people, 'alternative' or even 'environmental' lifestyles are characterized by thoughtful, healthy approaches towards *eating habits*. However, this is a misconception as food was not among the main interests of any of the households we visited.

Nevertheless, it is true that our respondents also differed from the rest of the population in their eating habits. It could be said of almost all of them that they ate very simply. There was an obvious inclination towards traditional Czech cuisine – or, to be more precise, the Czech cuisine of the poorer classes.[186] Meat was very limited, whereas cereals, legumes and various vegetables and potatoes had pride of place.[187] There was no evidence that modern-day dietary trends had influenced our respondents' appreciation of potatoes and traditional Bohemian and Moravian food.

There was a general tendency towards vegetarianism, but one which had a wide range: from a consistent approach ("people are squeamish

[185] 'Euro-Indians' is an endonym of people who try to bring elements of Native American wisdom, culture and lifestyle into their own and others' lives (Jehlička 2008).

[186] We will see how our respondents attempted to return to the lifestyles of their rural ancestors in other areas.

[187] In the summer, the Ryšavý family organized a social event – 'potato feast' – for their family and many friends. Prior to the new season, they made potato-cakes from the last of the old potatoes.

The 'Euro-Indian' Václav Franče tries to bring Native American wisdom, culture and lifestyle into his life.

about dead mice but are happy to eat a dead cow" – J.Ž.) to a mere preference for vegetables and bakery produce. This is emblematic of our Alternatives - their tolerance. There was not a single case of someone vehemently advocating vegetarianism or the 'right type' of diet. Nearly all of the respondents said they would not refuse meat if offered it when visiting someone.

We encountered a similar tolerance when it came to alcohol and smoking. Only two of our seventy respondents were teetotallers who would refuse to accept a drink. There was lower approval of smoking, though two members of our sample were smokers.

The reasons given for being vegetarian were mainly ethical and only occasionally health-related.[MK] An interesting answer was given by the South Bohemian I.S., a nature lover and conservationist: "Most people eat meat but are unable to kill animals. They don't want to get their hands dirty and see it as something disgusting. I admit that I kill animals and I've no real problem with it as I occasionally eat meat and I don't see why I should shove the responsibility onto someone else just because all that blood and stuff is not particularly to my liking... When someone kills a litter of kittens, they'll drown them, which is the worst thing you can do."

The majority of households we visited had at least a small garden, although often they were much larger. The decision to move to the countryside was frequently linked to the idea of extensive gardening or farming activities. Over time, however, reality began to set in: As people become involved with the local community and as they have to face up to crop failures (due in part to not using chemicals), then the majority of their plans are downscaled. There were, however, examples of the opposite phenomenon - over time the appetite for farming and gardening increased. Women in particular were often interested in herbs - during our visits we drank herbal tea and could choose from several types and combinations.

In terms of our research aims, the question of *clothing and fashion* proved to be deeply significant. We have already come across the Alternatives' approach to fashion in connection with the ownership of durable goods. There we saw that our respondents were immune to so-called moral obsolescence. This was also confirmed by the question concerning fashion in the narrow sense of the word – *i.e.* in relation to clothing.

Not one of our respondents was interested in the trends which fashion magazines might offer them. However, it should be noted that the

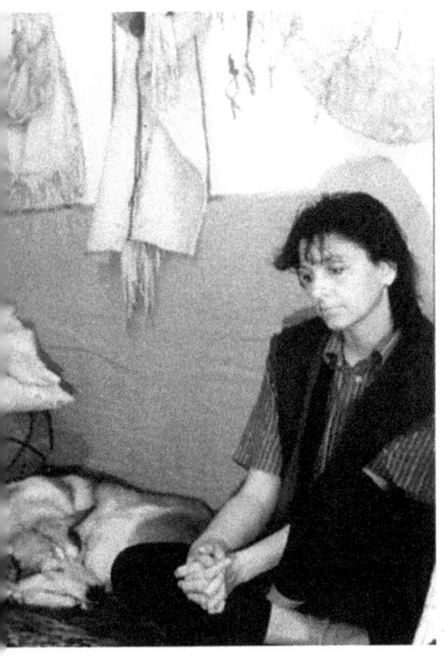

majority of women in our sample (as well as some men, *e.g.* from the subgroup of artists) were not indifferent to aesthetics in clothing. They had their own opinions, taste, style and expectations stemming from these. In order to dress themselves and their family well, they often relied on second-hand or charity shops, and even more frequently on sewing, mending and transforming their own clothes. These highly individual creations were then worn and mended over a long period. It became clear to us why the sewing machine was such an important item for the majority of Alternative households. There was another general characteristic evident amongst our respondents – a kind of faithfulness[UK] towards items of clothing: "If I become attached to something, I'll wear it till it's threadbare" – student A.P.

Some of the respondents even had a fondness for worn clothes. O.S. was rather extreme in his attitude: "I like the idea that some poor Kurdish refugee in Switzerland didn't want this horrible coat and now I'm wearing it."

However, we shouldn't oversimplify matters: There was a shadow of sadness hanging over some of the female 'Robinson Crusoes'. Isolation, no matter how idyllic it might be, is worlds away from the urban setting where it makes sense to show off a new coat. However, it's necessary to forgo that moment of pleasure in the interest of other joys. Mrs P., living in splendid isolation by a bend in the Vltava river, told us without bitterness and in the knowledge that you can't have everything: "I leave the house, the dog jumps on me and suddenly that's the end of fashion."

A natural aversion to wasting resources, the economically dictated need to save, and the presence of fewer appliances lead to a relatively low *consumption of water and electricity* in the household. However, we did not observe radical and systematic efforts by the Colourful to save energy or keep records of their consumption. There was an ongoing effort to do the right thing rather than a radical one-off decision or a revolution in housekeeping. A large degree of flexibility and tolerance towards oneself and members of the family was also in evidence here – "I don't want a tense atmosphere at home".

There is a similar approach to dealing with waste.[188] Most importantly, there is not much of it. Things are used until they are worn out; as we have seen, they are mended or given to others who might be able to use them. When dealing with rubbish, our respondents are among those who complain that there is nowhere to recycle their sorted waste. Often they became actively involved in setting up a recycling

[188] In general, the quantity and type of waste would be excellent indicators for research into lifestyles.

area in their local community. Some of them had piles of sorted paper, bags of textiles and plastic, balls made from compacted aluminium packaging and heaps of other metals and old bottles in their cellars and sheds, waiting in vain until the opportunity finally arose to get rid of them. Our respondents also differed from other village residents in one seemingly less important but illustrative detail: they did not burn plastic in their stoves – a popular source of fuel in Czech villages.

Recent years have shown ever more clearly the effect which *tourism* has on the world. Attempts have been made to transform hard tourism into a more environmentally and socially friendly soft tourism. It is an easy decision for our respondents – they stay at home and occasionally go 'for a wander with the kids'. The majority did not even feel the need to travel – at most going on some short trips around the Czech Republic or to Slovakia.

Only five or six respondents wanted to travel abroad – most often to the north (Scotland, Scandinavia). However, they too stayed at home, without any self-pity – they were resigned to their lack of means and putting their children's needs first. Two older respondents who used to enjoy travelling as students found that with advancing age they were more content to spend their holidays in the Czech Republic. Our research showed unanimous opposition to the traditional foreign seaside holiday, usually arranged through a travel agent. However, our respondents' attitudes should not be confused with soft tourism. In environmental terms they far surpass this, being closer to the radical demands of bioregionalism (Sale 1985).[189] This lifestyle notion of the Colourful directly contradicts the futurological idea which was formed in the 1930s of "a renewed nomadism by which mankind transcends its domiciliation and domestication" in the spirit of Buddha's teaching that "freedom lies in leaving the home" (Teige 1947, p. 16).

It is also possible to place *attitudes towards sexuality* within the wider context of consumption and modest lifestyles. In the West, awash with the atmosphere of 'sexual freedom', there has been a transformation within some groups which we might term a 'new prudery'. Although our respondents did not grow up in a society with 'liberated' sexual attitudes, they have responded in a way which might be said to be postmodern in spirit. However, this is more a trend towards traditional values, which has led them to a reticent or even critical view of 'sexual freedom'. Our respondents expressed the view that it was possible to counter the onslaught of a sexualized culture through emotional education. This view was most commonly held by women and

Christians. However, among the respondents there was also a high level of tolerance in this matter, along the lines of 'it's up to individuals, just as long as they don't hurt others'.

It could be argued that these are all restrictions, but when we asked our hosts about their overall feeling about life, we heard the reply: "I am happy." It was as though the people we had the opportunity to meet were reaffirming the idea of a rich life achieved through simple means.

'Simple in means, rich in ends'
– this is an idea we come across in virtually all cultures and epochs. It forms a counterweight to the human desire to have more and more. In the context of today's ecological crisis, it has taken on a new urgency and has become a watchword.[190]

Our suspicions are immediately aroused by the first clause, which is slightly disturbing. However, an environmentally friendly lifestyle need not only consist in a reduction of the basic necessities in life; on the contrary, it might see them flourish. But we could go even further and say that a rich life is not some kind of reward for heroic austerity. The process is in fact much less painful; if someone becomes interested in spiritual matters, then their interest in shopping and consumption disappears quite spontaneously. There was never any hint of self-sacrifice or the heroism of renunciation from our respondents.[MK]

It is impossible to give general instructions on how to live a rich life through simple means due to the fact that the picture provided by our research indicates a great variety in lifestyles. One thing, however, can be said with certainty: The values of the Alternatives not only lie outside of material needs, but also outside of themselves as individuals. Gabriel Marcel would say that their values were linked to a certain communion and that their self-realization is achieved through action rather than meditation. Marcel writes about this in connection with hope (1951).

However, these are quite unusual activities, especially in the context of contemporary Czech society. As has been shown through the vocational characteristics, and as we shall see again later, it is not about work which is financially rewarding, which is characterized by high-level performance and may bring victory in competition. Both during and outside of their work, our Alternatives carry out activities which in their eyes have a clear sense of purpose; specific work *for other people and for nature*.[OK]

[190] The names of D. Elgin (1981), B Devall (1988) and A. Næss (1990a) are associated with its formulation and elaboration.

[191] Rather than damaging it as a result of the stress caused by the lack of means, as we sometimes hear or read.

[192] Over the course of our interviews the women were the equal partners of their husbands.

The *family* is of great importance in the lives of these people. Being anchored in the family is characteristic of our sample – apart from the aforementioned group of older single people – and in many cases is the result of Christian beliefs.

From the interviews and the overall climate in the households, it is possible to conclude that a happy, secure marriage is one of the aspects where our sample differs strikingly from rest of the population. According to the respondents, living modestly at a slow pace benefits the family.[191] We have already mentioned that the families are bigger than average, and that several generations often live together harmoniously. Having a family is seen as a major, decisive step in life. This sense of responsibility[UK] may be one of the sources of the attitude towards pre-marital cohabitation. One young couple from our sample had chosen the premarital path of the 'new prudery' or the new trend of 'love without sex'. The partners had got engaged in the old-fashioned manner and although they share the same flat, they do not have sexual relations. They excitedly showed me their engagement rings – an old-fashioned attribute of marital vows and premarital purity. They believe that sexual abstinence enriches their lives.

As we know, the married women usually stay at home and look after the children. Does this mean that they are adhering to the traditional male and female roles? Perhaps it does, but there are several inconsistencies – for example, the activities and interests of the women are not confined to the family circle. As with the whole family, it is typical for them to be involved in the wider social circle, including public activities within the community. Neither has the patriarchal division of labour within the family impacted on the standing of the woman as joint decision-maker with regard to their lifestyle.[192] In many cases it even appeared that the woman were the ones who had taken the marriage and family on an alternative path through life.

We also saw challenges to the traditional ideas of masculinity and femininity on the part of the husbands. One of the most outgoing respondents, the environmental activist O.S., happily admitted that he had 'female hobbies' – "I enjoy cooking and flower arranging, I look after the kids and stuff". This unusual self-confidence is reminiscent of the 'alternative supermen', 'changing men', who approach the world with 'compassionate authority'.

One of the most striking features of the Alternatives is their *openness to wider society* and high level of altruism outside of the family.[PK] The family environment is not a sanctuary from the world; rather it is a safe base from which to go out into the world. This is fundamentally different from the supposedly 'happy family' which has to be laboriously built up and protected in the midst of this world, as in Franz Kafka's *The Burrow*, where it is isolated and anxiously fortified against everything that is happening outside.

Mrs R.V., who stays at home with her children, is selflessly engaged in the demanding task of establishing Waldorf nurseries and schools. One of the families from our contact list intentionally adopted a handicapped child. Another modestly living family has twice taken in refugees from the former Yugoslavia.

There is one particularly interesting observation relating to this: Our contact list of 'alternatively living families' contained several cases of children with severe physical or mental handicaps. Their parents did not respond to this challenging situation by shutting themselves off as a family. On the contrary – they were heavily involved in the life of the community and worked selflessly for other people. It is difficult to draw conclusions from this due to the small size of the sample and the lack of a psychological study. Nevertheless, it can be assumed that the presence of these families in our sample is not a coincidence and the need for the family to deal with the child's health issues leads its members to a more profound outlook on the world, to a reassessment of attitudes, and to the search for basic certainties and meaning in life.[PK]

A seemingly insignificant detail related to the character of Alternative families is their attitude towards pubs, restaurants, cafés and wine bars. It provides a good illustration of some of the features of our Alternatives and the contrast to the stereotypes people often have of them. None of them expressed any outrage at the idea of wine, beer, cigarettes or a lively night life. The pub is often associated with the idea of friends and pleasant company. It is a place you can happily crawl out to from your 'burrow'. And it is also a place where it's possible to get something practical done for the life of the town or village.

As we already know, interests and work activities of these people represent attempts to *reawaken life in local communities* (we saw their efforts in action in Miroslav, Olbramov, Valašské Klobouky, Volary and Žďár nad Sázavou). Although these activities are spontaneous in character, within the confines of a small town it is logical for them to be somehow linked to local-authority institutions. Among our respondents we met mayors, deputy mayors, councillors and members of various committees.[193] However, they had not been institutionally inspired to help the community – they were not 'carrying out orders from above'.

Magda Svobodová is a prominent figure in our research. She is a housewife and mother of three, but she is also a midwife, teacher and gardener, as well as a support and inspiration for her husband, a painter in Radňovice.

[193] However, we should also point out that in the autumn of 1992 many of them spoke with some bitterness about developments in their community and their own position. All too often they were coming up against the passiveness and selfishness of their fellow citizens. Most of them did not want to stand for re-election. They had also lost faith in 'big' politics.

We have already mentioned one railway worker: In recent years he has written and published the slim volumes *Pověsti šumavských horalů* (Legends of the Šumava Highlanders) and *Kniha o volarských božích mukách* (The Book of Volary Shrines), *the railway-themed book for children Pohádky ze šumavských lokálek* (Fairy Tales from Šumava's Trains) *and Volarské seníky* (Volary's Haylofts).[194] He takes photographs of events in the town and the surrounding countryside and is a chronicler of the town. In addition, he organizes an amateur dramatic society in Volary: "I was looking through old papers in lofts and cellars. Amongst a pile to be thrown out I found a list of German fiction on the Šumava region where they mentioned a German play about Volary. I started searching in archives and libraries. In the end a library in Prague sent it to me. It's beautifully written in Swabian. 'A drama on love and betrayal on the Golden Trail,' about 16th-century Anabaptists and salt-smuggling." R.K. talked 'a man from the neighbourhood' into translating the play. He then produced the play himself with Volary's amateur actors, and it was well received in the town.

Amateur dramatics also played a role in the renewal of life in Žďár nad Sázavou. One of our respondents, V.B., an engineer at the Žďas factory, belongs to a group of ecumenically orientated Catholics. One day, while sitting in a local restaurant, they suggested to their friends – pub regulars with completely different interests – that they put on the Passion Play at Easter. One can imagine what effect these months of intensive rehearsals and cooperation between people from different backgrounds had on the life of the town – not to mention the sell-out performances themselves.[195]

The third theatre initiative came from M.V., the mayor of the small village of Olbramov in the West Bohemian border region. She moved there from Prague a number of years ago. She has devoted years of her life to the village community through her unpaid position, whilst others have invested that time in furthering and protecting their own personal interests. She organizes children's drama and art groups in her own village and others in the vicinity. Although she doesn't consider herself to be a practising Catholic, she has made sure that many of the dilapidated chapels in and around the village have been restored and are now maintained. She has been behind a whole series of initiatives which go beyond her official remit.[196]

Mr and Mrs R. have helped to revive cultural and social life in Miroslav in South Moravia. They organise lectures, concerts and social events such as the U Floriánka fair. They have been pushing for the

[194] Note how the book titles relate to the local community and region.

[195] I saw a performance in Easter 1993. It was a grand affair which was important for Žďár but also far beyond. In May it was shown in Třebíč, České Budějovice and Prague-Břevnov for the celebrations of the monastery's 1,000th anniversary.

[196] Men performing civil service instead of military conscription have been involved in these activities. Other respondents also spoke of their help to the community.

town castle to be repaired and Mrs L.R. helps in the publication of the town newsletter. None of this is easy given the passive climate of today's small towns, in particular towns whose traditions have been disrupted – until the war Miroslav was predominantly German.

People who are involved in bringing life back to smaller communities are not always long-term residents or even their offspring. The simple idea of 'going back to your roots' is a naively romantic notion. The 'influx', as the incomers are disparagingly called, are often at the forefront of this renewal. It is often the local residents who can be found looking on passively at the incomers' attempts at improvements. "Mayor, your cement's gone hard," said a local Olbramov passer-by, watching as the mayor – formerly from Prague – worked hard on the village square together with a group of her Prague friends.[197]

This confirmed our expectations about the role of the village and the values attached to it as part of an environmentally friendly lifestyle. All of our respondents expressed a *more positive attitude towards the village than the town.* Although most of the Alternatives previously lived in the city, many of them emphasized with special pride how they had lost their city ways and now found it hard even being in a town. "It's tiring in Brno – even the little things like the cobbles, whereas now I'm used to walking on grass. When I'm walking along the main street I slink along the walls because there are so many people," said Mrs Ž., who lives in an isolated location below the Chřiby mountains.

Mrs J.S. from a South Bohemian farm: "I start to get aggressive when I've been in the city too long. If someone barges into me, I feel like shoving them back." And the 'Euro-Indian' V.F. from a nearby small town thought that: "In the city you're just waiting for death to come."

In the interview with Š's family, the son divulged that: "We wouldn't get Mum out of here anymore." The mother backed this up: "I'll give you an example – I leave the house in the morning, catch the bus and there's no-one else on it. When I imagine all that rushing about in the

← Amongst our respondents were mayors, deputy mayors, councillors and members of various committees (Ivo Stehlík).

↑ Train dispatcher Roman Kozák organises an amateur dramatic society in Volary.

[197] When I visited Olbramov for a second time, I heard some encouraging news: Evidently inspired by the successful drama activities of their grandchildren, four elderly residents of the community wanted to perform a production led by the mayor.

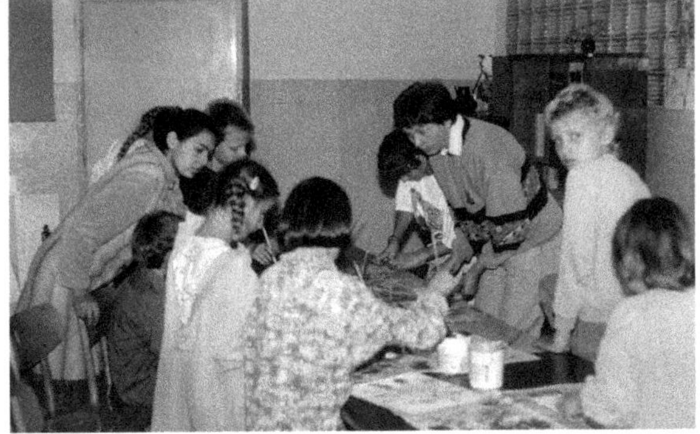

[198] Naturally, life in a village, particularly in seclusion, is much more difficult than in the city. It takes psychological and physical strength, as well as enthusiasm if you are to persevere with your decision.

[199] There has undoubtedly been a change in how the town and countryside are perceived, which has been demonstrated in sociological studies (Librová 1985). This does not just apply to a specific group of people. Blažek and Reichel (1991) discovered that more than 52% of city residents would like to move to the country.

underground and the stress of clocking in on time... that's a big difference... It's hard work here, but we don't have all that stress."

R.K.: "When I travelled to Prague to see my wife, I used to think how we'd be so cultured, but we'd always end up at the cinema on Wenceslas Square. Culture is more accessible in Volary."

Most of the respondents were living in a village or small town. Twelve of the families had left the comfortable conditions of the city to live in the countryside – often in complete seclusion. This was usually a gradual process: Weekend and holiday stays at the cottage gradually become longer; you have to water the ever-expanding flower beds, there are more and more rabbits and chickens, and eventually the decision is made. The people who made this choice had jobs and flats in the city. They did not leave the city to escape the housing crisis.[198]

However, the statements from our urban respondents were also interesting. They also displayed a great fondness for the countryside and were planning to move to the country in the near or more distant future.[199]

The countryside opens up another dimension in people's lives thanks to the *garden and domestic animals.* Working in a garden or in the fields is not only a source of partial or complete self-sufficiency in fruit and vegetables; it also gives people a new way of perceiving the world.[OK] There was only one case in seventy where the respondent said that she did not have a garden and was not interested in working with the soil.

The presence of a dog in the majority of Alternative households was a natural addition to the atmosphere. It was always considered a member of the family; the other households expressed the desire to have a dog, but gave their love of dogs as the reason why they didn't currently have one while living in a flat. These people at least spoke more generally about the relationship they had with someone else's dog – from the neighbourhood or when on holiday.

One special example of a relationship with the countryside and the soil was the decision to renew *farming on the family property.* I.Z.,

145

a zoologist and author of specialized publications, made just such a decision. He had plans that were highly ecologically beneficial: he wanted to grass over the land alongside a stream, encourage the formation of meanders and create baulks elsewhere. He was no dreamer – he was aware of the material difficulties faced by the private farmer in Czechia in 1992.[PK] This was why he spoke with restraint and with provisos. However, his work in the agricultural landscape makes sense today: "One interesting thing is that when we used to come here years ago, all of the fields were joined together. When we acquired this strip of land, we created two baulks by dividing the fields up. This led to the land being segmented and the crops being rotated. I had never seen partridges here before and then suddenly there were two of them leading out their young – with eighteen chicks apiece."

We visited married couples who had resolved to *renovate a farmstead* which had been returned to them through restitution – often a white elephant due to their terrible condition and the cost in terms of finance and labour. It was certainly not an improvement on the comforts of a perfectly inhabitable house. All of the Alternative builders were establishing basic living conditions or trying to restore the site to its original form. Z.Ž., the owner of a sawmill from restitution: "One person can't manage it alone... I'd like to use the money from the business to fix all of this and put it back to the way it was. We have a large chronicle in the form of my mother who remembers everything... Everything was perfect here before – ecological and economical at the same time... The use of water propulsion in combination with burning sawdust, a small turbine, no waste. It'll be impossible to have it exactly as it was, but that's our kind of model."

146

We were impressed to discover that after a tiring day these builders were able to lay down their trowels and find the time to do unpaid work for other people and for the community. It's also true that they were able to count on friends to help them with their work. Our respondents have a large number of *friendly and neighbourly contacts* where they live. In addition, those who moved from the city are often visited by their city friends, even though the visitors' initial enthusiasm for the attractive location and lifestyle sometimes wears off.

Another area of activity which fulfils people is *helping nature*. This is particularly obvious amongst our respondents from the Green subgroup. H.M. is an expert on, lover of and, most importantly, protector of owls and birds of prey who spends nearly all of his free time researching nesting sites. He is single and ploughs his earnings into it. P.H. also spends a large part of his business profits and almost all of his free time on raptor conservation and is particularly keen on hierofalcons. K.Z. chose a job which reflected his love of animals – he is a zootechnician who cares for ungulates in a zoo; he dedicates his free time to protecting tawny owls and other members of the owl family.[200]

Some of the Alternatives had arranged their lives in such a way that their *profession* reflected their priorities in life. They would then get caught up in their jobs – social work, nature conservation or art – and had to make a conscious effort to tear themselves away from it. Sometimes the opposite model applies: the job is a means of subsistence and the bulk of their real work takes place outside of it.

It would be wrong to assume, based on some of their other attitudes, that our respondents have a negative view of *private enterprise*. The majority of them are involved in businesses, though of a fairly unusual sort: sewing toys, carving wooden nativity scenes, small-scale production of roof tiles, alternative private farming, the sale of religious literature, running a family-owned sawmill, *etc.* These people don't share the usual goals of entrepreneurs: to get rich, invest more money in the business, make the big time and leave their mark. The only truly successful entrepreneur in our sample made money in order to protect birds of prey in Moravia.[MK]

So why then are our respondents running their own businesses? It would appear that they don't want to be dependent on institutions or have to be somewhere at an allotted time to 'clock in'.[201] These strong free-spirited people are naturally opposed to the idea of being controlled by someone else. In particular, they don't want to do anything of a vague or dubious nature. This is probably why so many of our group have tried to escape a 'secure existence'.

During our research we became convinced that attempts at 'alternative' lifestyles could fail if there was no independence in terms of property. We talked to one respondent whose alternative mountain farming had just come to an end – he had been evicted because the hunting lodge in which he lived had been sold off by the Forests of the Czech Republic in circumstances which were beyond his control.[202]

Our respondents were so immersed in their work for the community, nature and family that they didn't have any other interests which we would describe as hobbies. Their activities, such as organizing amateur theatre, making costumes for children's performances, helping to restore a church or even growing vegetables to supply a modest household, cannot be seen as traditional 'hobbies'. Meanwhile, stamp collecting, train sets or cactus growing would be too small-scale and superficial for these people, as they barely scratch the surface of human existence. When asked about their hobbies or small pleasures in life, one respondent answered with an apologetic smile: "I don't have small pleasures, only large ones".

This is also consistent with their (at best) lukewarm approach to sport. The majority of respondents saw sport as a waste of time or 'something I don't really get'. Rambling and canoeing were the exception. Six of the seventy respondents had previously been interested in sports, two of them at a competitive level. However, they all spoke about it in the past tense – they had given up competitive sport when they realized the demands it made on their time. In addition, some of the Alternative organizers of community life and nature protection had had bad experiences with sportspeople. When they went to a local sports club to ask for help with an event, they were usually met with incomprehension. The low number of sports enthusiasts is particularly striking when you consider how active the people in our sample are in social and group events.

Naturally, the activities we have mentioned enrich a person's life. However, they are an aspect of a person's inner life which remains largely hidden from the sociological observer. It is difficult to ask about them and difficult to answer. Quite often, even the person in question is unable to communicate their feelings. However, there are dimensions in which it is possible to express these feelings. One of those dimensions – one which forms the basis for the lives of our voluntarily modest respondents – is *religion.*

[202] He had a pre-emptive right of purchase which the seller did not respect. This was allegedly due to his strict approach to protecting the countryside, which clashed with the plans of the local mayor.

The research demonstrated that religious attitudes are firmly linked to this lifestyle, especially its modest dimension. Most importantly, nearly all of the respondents professed some type of faith.[TK] Only three of the seventy respondents said they were atheists.

Our sample included numerous believers and practising Christians from various denominational backgrounds. This classification, however, is only the backdrop of their faith – all of them had a highly developed ecumenical awareness and behaviour. Our respondents distanced themselves from a categorization which would isolate them from other Christian denominations, but also from other religions. Sometimes their faith is a kind of religious eclecticism, a unique synthesis of Christian and non-Christian beliefs and attitudes which these people have encountered during their spiritual quest. We were slightly surprised not to come across any true devotees of Eastern religions, although their influence was evident from the interviews.

The research confirmed that the social role of the parish priest was being renewed in small towns and villages, especially when tolerant, open-minded young clergymen stood behind the revival of social life. However, there was also mention of members of the clergy from the past who had made a bad impression on potential believers at a certain stage in their lives. In some interviews there was strong criticism of the practices of the church as an institution in the past and present, as well as the dark side of 'traditional' religion and the morals of practising Christians. Our sample consisted mainly of those who professed an idiosyncratic, rather vague belief in the existence of some higher principle. "The world has meaning. From the roots of the grass to the galaxies" (J.K.). Linking God with nature and coming to know God through nature were very characteristic features of the faith we encountered in our research.

In some cases there was a pantheistic understanding of the world: "I have literally deified nature" (K.Z). Or elsewhere: "To my mind, God may reside within a white hill or a stone" (M.V.). Elements of pantheism were also present in the faith of Christian-raised I.Z., who used to serve as an altar boy: "I believe that God is everything. All of nature and everything that surrounds us. And you don't have to go to church to be a good person. That's not a requirement."

At other times we heard mature statements exhibiting the certainty of the Christian conception of a Creator or Mover of the universe: "I'm convinced that it is possible to know God through nature. If you start to reflect a little, then you have to realise that there is not only Nature here with a capital N, but that there is a Creator here and that Love is here" (M.F.).

The authors of the 'greening church' concept would be pleased to hear echoes of a sensualist approach to faith and some elements of mysticism: "I go to church in the winter. In the summer I prefer to go into the garden on a Sunday morning. When I hear the bells ringing out from the basilica and then from the parish church from about eight till ten, then I'm more relaxed there than in church. I always sit there in the corner feeling bored... I'm uncomfortable with the order they have there: song–sermon–song–sermon" (M.Š.). It

was often reported that the aesthetic space of the church was important for their faith.

The conversation about religious belief showed that although the
respondents were by no means tempted by a technocratic or purely
rationalist view of the world, neither did they resort to a visionary or
fantastical approach. They generally expressed the opinion that mere
intuition is not enough,[UK] and that a general education in the humanities is an important component of personality. The aforementioned
'Euro-Indian', who had gradually disposed of the objects in his flat and
wanted to limit his possessions to things which were essential in life
and could fit into his tee-pee, was procrastinating over the final step
– the bookcase (which, incidentally, contained mainly history and art
books).

An indispensable part of the 'simple in means, rich in ends' attitude
towards life is *art*. Both the children and adults in these households
take an active part in music; musical instruments and sound systems
are integral parts of otherwise modestly furnished flats. Folk is a very
popular genre, followed by classical music. The private farmer D.M.
found time in his busy schedule to sing in an Protestant choir in his
village. The paintings on the walls of the flats were usually reproduc-
tions by world-famous artists, though there were often originals as
well. Most of these were artworks by the residents themselves or their
friends. From our perspective, it is unimportant whether these were
amateur efforts or the gallery pieces which we saw in the flats of the
three professional artists.

The ability to appreciate art is closely linked to another gift: the ability to see grace and beauty in small things and seemingly everyday
moments, and to derive joy from them. When A.P. said to us, "Every
small thing brings me joy," she was not contradicting what we heard
earlier, which at first sight may seem to be the opposite: "I don't have
small pleasures, only large ones."

It is not easy during an interview for the sociologist to ask a respondent a general question about their attitudes toward other people and
whether they feel happy or unhappy in the world. Our hosts, though,
were not offended by our intrusiveness and were happy to answer.
With one exception everyone admitted – some spontaneously, others
more cautiously – that despite not always having had positive experiences, they would always come back to trusting people.[203]

Although the question concerning the feeling of happiness did
not provoke the exaggerated reaction of exhilaration familiar from

The teacher Miloslava Figarová: "If
you start to reflect a little, then you
have to realize that there is not only
Nature here with a capital N, but
that there is a Creator here and that
Love is here."

[203] A very different picture was given
by the international representative
survey *European Values Study* (EVS
1991). It had a sample of 2,109 people,
with 25% of the Czech population stating that it was possible to trust other
people while 72% said that it was necessary to be careful.

150

members of charismatic sects, there were no feelings of unhappiness in life (with one exception).[204]

"I'm happy, but I'm aware of the strange undertone of environmental pressures within this feeling of happiness," V.R. said. D.J., the deputy mayor of a small South Bohemian town, saw the limitations to his happiness in other ways: "If my table is laden but someone else is suffering, then I lose my appetite" – a statement which seems to sum up A. Næss's universalism, according to which "we should not live on a material level we cannot seriously wish others to reach" (1990b).

"I'm a happy person," said the mother of the severely handicapped girl, "there are just some times when... But I feel as if I'm on the right track in life."

We will conclude this chapter with another particularly Colourful quote from one of the Greens: "I'm more likely to say happy because it's a question of quality and quantity. The moments of happiness are rarer, but they're so intense that on balance I'd say they outweigh the others. There are moments when I'm happy to be in this world, to have the chance to look at the stars, to see that diversity [in nature – H.L.]. When I add it all together, then I can see just how breath-taking it is. (...) One of the greatest human aptitudes is the ability to admire" (M.J.).

Reasons and sources

Answering the question WHY is always difficult. This was also the case when we asked why people voluntarily chose a modest life, as it turns out there is no common denominator for its source; it is not based on a single idea or even on direct imitation. We can also rule out any immediate inspiration from a film, television programme or book. The people in our sample do not adhere to any ideology (unless the ideas of the Christian religion can be considered as such), even though they might recall having been influenced by yoga, macrobiotics or Eastern philosophy. They are not even motivated by trying to be distinct from others – which is typical of sects. It is illustrative that after my initial experiences I had to remove formulations such as 'you live differently from other people' from the interview guide. There was no indication that our respondents wanted to live in uncomfortable conditions in order to test the limits of their capability, resilience and endurance – a reason which plays a role for those who go to live in remote, inhospitable regions.[205]

Their lifestyle choice has been *formulated individually*. It is the authentic and creative outcome of mental work. In the main it has emerged from the internal set-up of the personality and is rooted in deep psychological structures.[MK 206] The respondents themselves feel that they can't really take credit for their decision. A kind of fatalistic approach was characteristic – "I think it's my path."[PK]

For that matter, we can only describe this very loosely as a *decision* to change one's lifestyle. More often we observed a gradual transformation or crystallization resulting from specific conditions and events in life. This is an "ethical reflection on the direction of one's life" which is not even rationally justified (Rozbořil 1992). It also transpired that it was wrong to talk about 'intentional modesty'. As we have shown, the modesty of the people in our sample was not their intention. It is more appropriate to use the term *'voluntary modesty'*.

T.R., a selfless organizer of small-town cultural life, even said with a kind of self-deprecation: "We don't have a talent for it [fields, farming, the home, herding – H.L.]. Maybe we're involved in culture because we're lazy and we don't want to hang around the fields and the house."

This is related to the characteristic of *freedom* and *flexibility* in their approach to life which we repeatedly encountered in our interviews and is another feature which is distinct from an 'alternative' lifestyle based on the duties and rules set by communities or the ideologies of sects and movements.[207] Our Alternatives are extremely tolerant of each other and those around them: In response to my question, 'Why don't you have a car?' the slightly taken-aback J.S. said: "I don't know... We don't have one... I'm not saying we'll never have one, but we don't have one right now..."

Václav Bělohradský (1990) wrote about *ethically responsible* people, who are guided solely by an inner sense of duty, and *ethically convinced* people, who are capable of promoting their own view at any cost. Our Alternatives are distinctive in this respect – they do not profess any convictions, only their own responsibility towards people and the world. This is not to say that modestly living people have no coherent worldview. On the contrary, almost all of our respondents thought deeply about their lives, both on a general and a practical level, particularly in the context of tragedies and dangers in the world. However, this personal philosophy was only loosely connected to their lifestyle as an overall attitude towards life. It is perhaps also typical that in most cases there were no references to theoretical standpoints expressed by philosophers or other authorities.[208]

However, this way of life does not stem from any deliberate attempt at 'self-realization', if this is understood to mean a focus on one's own

[206] As has already been said, this is probably a key factor providing an opportunity for these types of lifestyles to endure.

[207] In the history of Europe and other cultures we could find dozens or even hundreds of such movements and programmes – for example, the attempts by the Czech interwar avant-garde to introduce a social, health and architectural programme for a new lifestyle which minimalized people's needs.

[208] A psychologist would look for the motivations of the Alternatives in their strongly developed emotionality, which is outwardly manifested, for example, in their interest in art.

personal development and liberation from ties to society. In the 1960s, thousands of young people in the West sought this kind of self-realization, but the liberation from obligations also brought with it the burden of anxiety and uncertainty.

Today's 'alternative', which our research is concerned with, is not 'civic escape'.[209] It is clearly a *socially committed lifestyle.*[OK] Freedom is directed towards an interest in the surrounding world, to helping people and nature. The Alternatives' calm and self-evident position, lacking a rigid programme, is in fact "the prolongation into the unknown of an activity which is central – that is to say, rooted in being" (Marcel 1969, p. 20).

Somewhat surprisingly, and contrary to Ronald Inglehart's conclusions (1977), it seems that the *limited material possibilities* of young people and young married couples can sometimes be the 'trigger' for a more spiritual way of life. There is a certain type of rationalization in operation here – literally 'making a virtue of necessity'. Once a modest mode of life has been adopted, its adherents then remain true to it (particularly if it is supported by the views of important people in the area or by information about trends in the West), even if they could 'improve' their material situation (*e.g.* by completing their studies, or once their children have grown up). Therefore, the willingness to live modestly need not only be the consequence of resentment towards overabundance.

Judging by the interviews, it is likely that the sources of 'alternative' attitudes and lifestyles can be found in *childhood*. It is interesting that the respondents did not think that their parents had influenced their choice of lifestyle (at least in the most numerous age group – the thirtysomethings). The *grandparents* are mentioned most often in connection with the formative events of their childhood, in particular grandparents living in the countryside. In their eyes, it was Grandma and Granddad who taught them the values and skills which they cherish today. They played the role of companion, expert, authority and leader. V.F., the South Bohemian 'Euro-Indian', told us: "Granddad liked nature – in fact, he loved it. (...) Granddad had a great gift: he was able to talk to trees. Of course, I didn't realize this right away. (...) Shortly before he died, Granddad walked around the garden and went up to an apple tree and said: 'My girl, we'll be going together.'"

Not as mystical, but just as vivid, were the memories of H.M., the raptor conservationist from Březolupy in Moravian Slovakia: "Whenever I had a problem, I'd always go running to Granddad. I helped him when he brought logs from the forest, I helped him cut wood with the saw, he taught me how to make birch brooms, he taught me to do loads of things that were related to nature and farming, and I enjoyed it so much. Usually when my folks were looking for me in the evening, they'd find me with Granddad, even if I'd been told not to go there."

V.M., a nature conservationist from Slavičín, recalled how his grandmother got angry with him when he and some other children trapped butterflies in boxes: "You wee beasts, let them go! The Lord God created them and there you go catching them."[TK]

Older respondents (50–60 years old) tended to remember their own parents in a similar way. However, the context of interviews does not support the conclusion that this contradictory case was an example of 'late obedience' (Lorenz 1974, p. 69), compensating for earlier intergenerational conflict. It would appear that a correct interpretation of the influence of parents and grandparents needs to focus on historical rather than psychological roots – on the different experiences of these age groups: The critical attitude the majority of the young Alternatives have towards their parents reflects the sad plight of Czechs now in their fifties and sixties who spent a large part of their life in a society which followed the deluded path of communism.

However, there is almost certainly a more general reason which transcends the country's political history: The fathers and mothers of today's thirtysomethings represent the world of modernity (the fathers became cogs in the wheel of industry and merged the fields together with heavy machinery, while the mothers scoured the shops for goods), whereas the world of their grandfathers and grandmothers has come to symbolize a (certainly idealized) world of tradition where the purpose of existence is clear.

In addition to the theme of 'Grandma and Granddad', when searching for lifestyle sources the interviews were dominated by memories of the place where the respondents grew up – of an event or *story from childhood tied to a specific landscape*. The story, the landscape and the grandparents were often interlinked. Many of the Alternatives, in particular the Greens, also grew up in a landscape influenced by a Wild West, back-to-nature culture which had been popular in the Czech Republic since the 1920s, known as *tramping* (Jehlička and Kurtz 2013). One eloquent respondent spoke of the 'tramping bypass'. In spite of the differences in the motivations for an environmentally

friendly lifestyle, we can say one thing with certainty: None of our respondents grew up without any contact with nature!

At the same time, though, it transpired – at first sight illogically – that *they are people with long-term experience of an urban environment, usually a city*: they have lived and studied in towns and cities. The ability to appreciate a rural community and landscape in contrast to a town is one of the fundamental sources of their current attitude to life.[210]

On several occasions during our interviews with married couples we were given the impression that it was the woman who had set the course for the family's lifestyle. The sociology of the family supports this hypothesis; several studies have confirmed the old metaphor that the woman in the family is the 'neck which moves and turns the head'. For example, among our married couples it was usually the woman who had initiated the move from the town to the countryside and it was usually the woman who was inclined towards the religious faith which underpinned the 'alternative' lifestyle.[211]

However, there is a question mark over how long the women will persevere with their decision to pursue an 'alternative' lifestyle. Much more so than men, women are prone to the feeling that a life in modest circumstances blocks 'other' opportunities in life. The attitudes of many women exhibit certain 'romantic' features which can lead to them initiating the 'alternative' lifestyle but are also more susceptible to disillusionment in connection with its realization.[212]

It is interesting that *systematically looking after one's health* did not feature as a motivation in our sample. This was probably no coincidence; nor can it be explained by the respondents' youth or the fact that the interviews did not turn to issues of health and healthy living. The fact that these people live in a relatively healthy way is more a consequence than a motivation.[MK 213] However, concern for the children's health in the family did play a role – for example, when deciding to relocate. Some of the families which had moved from the town to the countryside stated that one reason for moving was that their children frequently got sick in the town.

[210] There is a similar 'urban' basis to people's emotional attachment to the countryside in a historical analysis (Librová 1988).

[211] On other occasions I met the wives of budding entrepreneurs who were disturbed by their husband's sudden change in values. Blažek and Reichel (1991, p. 18) also found a higher level of spirituality among women. However, this 'lifestyle-forming' role of women can also work in the opposite direction; the woman is against living in uncomfortable conditions and thwarts or dilutes her partner's initiative. The trouble young farmers have finding a partner is well known, then there is the notorious woman portrayed in literature who pushes her husband to better his career, earn more money and live ostentatiously.

[212] The problems young farmers have finding a partner are typical from this perspective. The difficult and unromantic character of their work is clear from the outset.

[213] This is similar to the issue of eating habits: Many people, including journalists, wrongly equate an 'environmentally friendly lifestyle' with looking after one's physical health.

Unfortunately, even shamefully, our observations showed *schools*, in particular primary schools, in a very negative, if not downright shameful light. From our sample only two respondents thought of their teachers and the primary school as something which influenced them in life. What was even more interesting was that both of the women who spoke fondly of their school went to a one-room village school! "I have great memories of the one-room school – it was something I would love my children or grandchildren to experience." When our respondents did recall being inspired by an institution they had attended as children, it was not the teacher in school who came to mind, but – in addition to 'old tramps' [in the Czech *tramping* sense – translator's note] – the leaders of various after-school clubs.

There was less scepticism when it came to higher education. Interestingly, our respondents often remembered their *art teacher.* We might ask whether this was coincidental or due to the fact that at secondary school art was one of the few subjects representing a cultural and emotional education. It is worth recalling that as far as tertiary education is concerned, many of our respondents who were studying technical or scientific disciplines left after their first year of study.[214]

One significant motivational factor amongst our respondents was, of course, their *attitude towards religion*,[TK] as discussed in the previous chapter.

An interesting role was played in the life of the Alternatives by the *political changes after 1989* [following the Velvet Revolution in November which led to the fall of the communist regime – translator's note]. This is understandable as our respondents are generally very socially committed individuals, many of whom actively took part in the events of November 1989 ("Who else is going to climb up on that fountain?!" said M.J.'s fellow citizens at the time, spurring him on). Even before the Velvet Revolution, some of our respondents had been involved in overt or covert opposition to the regime (*e.g.* distributing samizdat publications in small towns). Many of these informal leaders from the pre-revolution days went on to be elected as councillors in local elections.

However, neither the events of 1989 nor the subsequent changes in society were the source of our respondents' environmentally friendly lifestyle. Likewise, the restitution of property, which we touched on earlier, was only a means and opportunity to realize their plans.

The environmentally orientated reader will still be searching in vain for an answer to the question: What role do an environmental

[214] This selective abandonment of science and technical colleges is potentially dangerous: People who are sensitive towards non-consumerist values thus leave the field clear for pure technocrats.

conscience and environmental education have in the establishment of an environmentally friendly, modest lifestyle? We shall now attempt to answer that question.

How much 'green' is there among the Colourful (and vice versa)?
The main protagonists in our research into an environmentally friendly lifestyle were not members of the 'green' movement. We were primarily interested in the Colourful Alternatives. They represent a great cultural diversity with their life stories, interests and activities. They don't have many features in common – the basic one is our selection criterion of *environmentally friendly modesty.*

However, not even the environmentally friendly aspect of the Colourful's lives has been programmed in a clear-cut way, *e.g.* environmentally. The willingness to live modestly is not rooted in environmental awareness, but once again in a variety of different reasons. It is a secondary *unintended consequence of a life heading in various directions, but never in the direction of shopping and consumption.*

But be careful: this does not mean that the Colourful are indifferent to nature. They have a lively and profound relationship to it, though even this is more of a by-product of their overall direction in life.[215]

So, if the Colourful are not directly environmentally motivated, are they even aware of the environmental consequences of their lifestyle? Even here the results are not completely clear. Some of the respondents replied immediately and convincingly explained how their lifestyle was environmentally friendly. Other respondents were more hesitant in their replies – with some it was obvious that they had never thought of such things before the interview, while others expressed the reticent opinion that their modest life had such an insignificant impact in terms of the ecological crisis that there was no point in even talking about it.[PK]

However, we did establish that all of the respondents were *relatively well informed about environmental issues,* though most of them did not actively seek out ecological literature. If they happened to come across an environmentally themed article in a newspaper or hear a programme on the radio or television, they would follow it with interest. Our respondents' spontaneous and intuitive approach to the world meant they were able to grasp ecological issues without having wider specialist knowledge. To quote our respondents: "In my opinion, these things are simple and you don't have to be a genius to understand it. If you're going to build some kind of monstrosity, then there are obviously going to be repercussions – the only question is what they'll be" (R.K.). "I don't keep up with environmental literature, but I do follow every tree that disappears from Brno. I'm aware of the withered rows of trees whenever I drive through the South Moravian countryside, the withered windbreaks..." (V.P.).

The prominent Czech painter and graphic designer Jaroslav Šerých has a mature environmental philosophy which is in line with the urgent recommendations found in environmental ethics journals. "I'm not going to put those horrible substances into the soil just so I can grow bigger radishes! Their quintessential feature is not in their size."

But it is not just a question of the Colourful living ecologically due to their modesty and being aware of nature and its fate – they also spontaneously *act*[OK] on its behalf. If trees are being planted in or around the village, then it is often our Colourful who are the only citizens who come to help. For two years in a row, the artist Mrs Č. and her son volunteered for the forestry work in the area around their home. The reasons she gives are ecological: She is concerned by the state of the forest and by the lack of moisture in the soil. She would like to contribute to reforestation. Another example of the spontaneous, practical environmental behaviour of the Colourful is the way they deal with waste, which we have already discussed.

I consider it significant that a large number of Colourful Alternatives are directly or indirectly involved in *local decision-making* and influence it greatly in a pro-environmental sense. Our respondents from Volary – civically and culturally engaged members of the council – pushed for the council to adopt an ecologically based stance regarding the completion of the Temelín nuclear power station, and have been making efforts to construct a wastewater treatment plant and safe landfill sites, and have campaigned against unwanted interventions in the surrounding countryside, and so on. In the small village of Olbramov, it is thanks to the Colourful mayor that household waste has been sorted and recycled for many years.

We wrote about the 'green' character of the Colourful's activities when renovating their old farms and the adjacent land. In this respect, their ecological contribution is apparent to everyone.

However, in terms of environmentally decisive changes to society, what is even more important is the impact the Colourful have on small communities, which neither they nor the casual observer are even aware of: Through their attempts to bring about a renaissance in the life of villages and small towns, they are implementing a political and

The painter Jaroslav Šerých: "Nature cannot be defended solely by attacking because it is also an internal matter."

216 In his study on the sociological aspects of the Nové Mlýny reservoirs, Libor Musil (1992) highlighted how closed off the environmental activists were and how unable or unwilling they were to establish contact with the wider social milieu. The activists who wanted to drain the dam couldn't find an appropriate way to approach the local inhabitants whose lives would be directly affected by changes to the landscape and who would ultimately affect the outcome of the conservationists' plans.

217 No doubt this was because our informers recommended people who were approachable and communicative. My own experience and observation of nature conservationists confirm that this type of conservationist exists and is relatively common.

sociological idea which sees the rise and development of the communal autonomy and cultural identity of regions as one of the key conditions for an environmental breakthrough.

Throughout the research, we watched with avid interest to see how the answers from the subgroups of the Colourful and the Green would differ. The initial hypothesis about the adaptability of the Colourful in contrast to the Greens' lack of versatility was only partially confirmed.

In terms of our criteria, the group of nature conservationists was further divided into two typological subgroups (with, of course, the existence of crossovers). The first group was strikingly similar to the Colourful members of our sample. They are people who are clearly linked in many ways to the life of the local communities, who are open to social contact and communicate in an outgoing manner. They have other features in common with the Colourful which are described in this text: A desire to follow family traditions (in particular the world of their grandparents), strong religious sentiments, an interest in art, tolerance in their attitudes to those around them and to their own way of life, the ability to trust other people, and a subjectively experienced sense of happiness. We often find these Greens on local councils and in other institutions. We found an outstanding representative of this type, A.H., in the Prague Mothers environmental action group, and who was obviously not the only one there.

The second type of nature conservationists are people who have these qualities in a very weak form or not at all. These Greens are completely closed off in their own groups and are loath to leave their own bubble – "Most of the people I know are animal lovers. I don't really get on with people who aren't. I don't avoid them, we just don't have anything to say to each other." When I asked why two active groups of people in a small town – the amateur actors and the nature conservationists – didn't join forces, the expert on local life in Volary, M.J., told me: "I could probably find some actors who were interested in nature, but the other way round would be much harder."216

Amongst the Greens it is also possible to find people for whom nature conservation is connected to a hostile attitude towards people and even, paradoxically, towards the world in general. "I don't like people," "I take a dim view of human society," said some of the participants at a FALCO event interviewed by M. Misíková for her student research (1994).

We only came across one such young person in our subgroup of nature conservationists.217 He generalized about people being evil, though he exempted his own friends from this and praised them as "good people". His (perhaps slightly stylized) misanthropy was also directed towards art because he was convinced that it manipulated people and he wanted "to stay as I am". He "didn't care if God exists or not" and thought "the church makes fools of people". His answer to my question about happiness – "I'm unhappy" – was in marked contrast to the quiet satisfaction of the Colourful.

Should we be surprised to witness the helpless anger and desperation of nature conservationists who devote their attention, interest and feelings to saving what's left of the living world? Can we blame them for not wanting to compromise and for feeling that they only understand one another within their 'green' groups? Can I reproach them for their feelings of mistrust towards people who get in their cars and demand that more and more motorways be built? For their hostility towards nest robbers and heartless livestock transporters, their hatred of rich Italians who come to our forests to shoot warblers and tree pipits? I don't reproach them. There are moments when I understand them with all my heart and soul. But at the same time, I know that anger and resentment are bad counsellors[PK] and signify the true loss of hope, which is a crucial thing for us.

A conversation about the end of the world

When, as a researcher, I sat face to face with my atypical, distinctive hosts, I wasn't worried that they'd be fazed by the question of how they saw the future of the world – the planet, nature and humanity. However, I didn't expect to touch on a topic which had been so profoundly reflected on, often with excellent arguments.[PK]

Many of them began their answer with the words, "I've thought long and hard about this." But even without this opening, the tone and content of their answers were convincing: Only one respondent expressed the conviction that "humanity will somehow fix it". Another (by coincidence, the only successful entrepreneur in the group) stated that he didn't think about this issue – that he didn't want to think about it. Otherwise, all of the respondents expressed the view that the situation was catastrophic, if not beyond saving. I came across attitudes which were very much in line with Marcel's concept of desperation as the precondition for hope. Many people had been struck by fear and a feeling that a line had been crossed from which it would be impossible to return when they read about the expanding hole in the ozone layer. It is curious that the Colourful were even more sceptical about the state of the planet than the active nature conservationists and ecologists.

The conversation even turned to the end of the world (Jiří Kostúr).

"Humanity is heading for disaster... It is on the edge of an abyss. It is caught up in the pursuit of technology and money. I see it as an image of the apocalypse, people rushing around, everyone grabbing what they can. They don't think about the fact that this life is just a kind of prelude, that eternity lies elsewhere, that the meaning of existence lies in something else. (...) People take it for granted that there are crops, water, sun – 'Mother Earth, provide, for I am the Lord here!' Man is the only creature on Earth which harms it. One day he's going to pay for it and it won't be long now." This was the view of the language teacher J.Ž., who is not a committed environmental activist.

"We produce so many things we don't need. The sheer demand for energy... I don't believe in a systemic change because of people's orientation in a consumer society. And it's impossible to change. It can be regulated to a certain degree, but I'm a sceptic. (...) Some kind of force would have to be used against the majority... and that's a dead end because it would mean violence. (...) It might be addressed by small groups of people, individuals who are prepared to live a different kind of lifestyle and protect their territory. But that would mean conflict and organization too, some kind of agencies and so on. But to change things fundamentally is impossible. I have to deal with it on my own, for myself" (forestry worker J.K.).

"In our family it can be seen in the way the children are brought up, to show them something different. But it's not going to play a major role in saving the world. It's more of an attempt at a personal solution. We want our children to be aware of the consequences of their actions" (Mr and Mrs K.)

The pessimistic views of the Christians in our group were moderated somewhat. They spoke of the existence of hope based on a belief in divine intervention or help. "He who built the pillars of this world will also look after it. So I don't worry about the idea of the end of the world, even though there are very serious signs that a different era is coming, that it is the end of this time" (scientific researcher V.P.).

Or elsewhere, referring poetically to the writer Petr Chudožilov: "Angels hold the World in balance, and if it tilts too far (to the wrong side), then they will quickly rise up on their tiptoes..." (H.B., a preacher from the Hussite Church, now a housewife).

"In technological terms, there's no way back, and there probably isn't a way to reverse it in political terms either, but I believe there is a way... by turning to God and living according to the Gospel" (retired teacher M.F.).

The physicist V.M. said at first quite vaguely: "It's mankind that is most at risk; the planet won't be destroyed. I feel that the world is different from how we perceive it. That it is different in terms of time, space and matter. That some things only appear to be the way they are... Our existence is unclear. I have to try to make sure that things don't go badly for mankind, but on the other hand, philosophically speaking, I don't know what to do... It all adds up to something quite unexpected."

The South Bohemian 'Euro-Indians' are more straightforward and determined. They are actively making preparations to survive the ecological collapse through foraging and simple farming. They are practising the art of living in the wild without civilization. The woman carries a scaled-down household on her back – "a kitchen and blankets" – and the man a tee-pee. They say they haven't got very far yet – they know how to make leather without chemical agents. They still go to the shop to buy supplies when they are camping. The 'Euro-Indians' do not deserve to be mocked. They have taken the idea that their lifestyle has to change seriously and they are being more conscientious about it than the rest of us.

IV.2. A report on the Colourful ten years on

(Reprinted from the book *The Half-Hearted and the Hesitant: Chapters on Ecological Luxury* [2003]. H. Librová has made only minor changes, adding some indices referring to the theoretical keys.)

During the research interviews in 1992 and then again during discussions about the book *The Colourful and the Green*, there was one question that kept cropping up: how the Colourful were likely to be affected by the course of time, the passage of the years. The idea gradually took shape that I should ask the Colourful if I could revisit them ten years on, whether they would be willing to receive me. Of the 47 families I had spoken to in 1992, I selected fifteen of the most typical.[218]

Researchers who wish to repeat a study after a lengthy period of time has elapsed tend to have a problem during the preparation stage. They can't be sure whether the old contact details will still be of use to them. I was able to get through in every case. This was the first valuable piece of information: the Colourful are still living in the same place as they were ten years ago.

When I asked my former respondents for permission to visit them, none of them refused. Their manner on the telephone was familiar. My Colourful answered me cheerfully and in the approachable way I had found so characteristic of them during my previous visit. The feeling of nothing having changed was reinforced by their declarations that they wouldn't be suitable for a follow-up interview because ten years on everything was "the same as ever" in their households.[219]

Could this really be true? When I was drafting the interview guide, I was aware of the range and intensity of possible changes. Over those years, time had passed in people's lives, the children had grown up, some of the Colourful had almost reached the zenith of their lives, and the parents they shared a home with had grown old. However, the Colourful might also have been beset by unpredictable events, illnesses or even death.

I had some knowledge of the latter. In 1994 I went to Tomáš Ryšavý's funeral. One late November afternoon, I said my goodbyes to him in the crowded Protestant church in his home town of Miroslav, where he had been a key figure in civic, religious and cultural life. During the preparation of the study I learned that the deputy mayor of Velešín, Dalibor Janák, had died.

Given all this, could the lives of the Colourful have 'stayed put in moving sands'?[220] Could they have been spared the changes the whole

[218] This time I asked the respondents for permission to publish their names.

[219] Others expressed concern that they would disappoint me: "We're no longer the Colourful we used to be." I became aware of a similar reticence in self-evaluation when I was preparing the first study. Back then, they had claimed: "There's nothing special about us, we just live the way everybody else does."

[220] The question is derived from a phrase used by Zygmunt Bauman to emphasize that the individual is unable to resist the effects that a rapidly changing society has on their life: "One cannot 'stay put' in moving sands" (Bauman 1998, p. 78).

of Czech society had undergone? After all, these changes have affected everything we can think of. In 1992 I captured their beginnings in the research, but it was only in the years that followed that these changes fully manifested themselves in our lives: The transition to a market economy altered the nature of employment and hit job opportunities hard. Those who were bold enough to do so were able to start a business of their own. Freedom has influenced our lives for better or for worse. The situation in schools changed, as did the role of religion and the church. Democratic norms began to be established: every citizen can have a say in public affairs and be elected to various offices and positions. Price liberalization radically altered the circumstances of household management and consumption. People's lifestyles and attitudes have also undergone a transformation. Consumerist competitiveness has increasingly interfered with neighbourly relations and the school classroom. Advertising has crept into the media, our streets and the landscape. It pesters us and our children.

In the decade I'm concerned with, something else happened that only the doomsayers had prophesied: The post-revolution enthusiasm gave way to disillusionment, and many good intentions and efforts began to be eroded by apathy. Idealists came to feel that they had been deceived; those of them who didn't wish to appear naïve shrugged it off, began to look out for themselves and bought themselves a holiday in a luxury resort.

In September 2002 we set out for the first visits accompanied by documentary film-makers. My head and my papers were filled with questions, but most of them shared the same subtext: Will I really find the Colourful, their households and life stories, their attitudes and opinions, as little changed as their addresses and their friendly manner?

No more moving house. A different view of the women

As I had gathered from the telephone calls, only one of the fifteen families selected had relocated from their town or village in the past ten years.[221] And they hadn't gone far – only about 20 km. In 1992 the Plojhar family had been living a secluded life on the Podhradský farm in a bend of the River Vltava. However, they had to leave because they were unable to buy the surrounding land, which was essential to their plan of building a pub with accommodation for paddlers. However, they hadn't sold the farm buildings but were renting them out to "people who want to try living in a remote area". They are now living in close proximity to the city of České Budějovice in a spacious

[221] Not counting one man who had moved out following a divorce, whose family continue to live in the family home, and one other family which had moved from a block of high-rise flats in Volary to a country house in the same town.

The 2002 research.

[222] In the first half of the 1990s, the rate of internal migration (the number of people moving per thousand inhabitants) ranged between 25 and 20, and the trend was a downward one. In 2000 the figure was 19.4. The issue of migration by the Czech Republic's inhabitants and their motivations for it was something I dealt with in the article *Decentralizace osídlení – vize a realita* (The Decentralization of Settlements – Vision and Reality, Librová 1996, 1997). The low migration rate of Czechs is often interpreted as a consequence of the problem with the housing market in the cities. In the case of the Colourful, who predominantly live in the countryside, this does not apply.

[223] Unfortunately, I did not have the opportunity to talk to them.

and well-equipped urban-style house which they started building immediately after their wedding, before moving to the Podhradský farm. However, they have not ruled out returning to the remote setting one day. The children miss the sledging and skiing which was practically on their doorstep. The parents miss the view of the valley and the sound of the river, but most of all – due to the light pollution above České Budějovice – the stars.

The low migration rate of Bohemians and Moravians has been noted by statisticians and remains curiously unaltered by the dramatic economic and social changes.[222] It shouldn't come as a surprise in the case of the Colourful either. However, it should be noted that in 1992, they might have come across as eternal seekers, as a migratory, restless element, compared to the average settled Czech citizen. At the time many of them had recently moved from the city to a village or a small town a considerable distance away. The move was part of the radical change of lifestyle they underwent. As it transpired, it was not to be repeated. They were to remain faithful to their place of residence.[UK]

One of the most conspicuous changes in the lives of the Colourful is the fact that the women have found jobs, some of them after getting a university degree through distance learning. While in 1992 they were all stay-at-home mothers, they are now employed; for example, as a teacher, a nurse, a textile artist and an entrepreneur/graphic designer. One of my female respondents works on projects creating halfway houses for people coming out of prison and also works with prisoners as part of a chaplaincy programme.

As the children grew up, the female Colourful joined the ranks of women attempting to combine looking after their family with other work. They identify with their jobs and enjoy them. This does not tally with what I wrote ten years ago. At that time I emphasized that the families of the Colourful were characterized by the traditional division of male and female roles. Have the Colourful changed in this respect? It is also possible that I did not observe carefully enough back then, that I relied too much on external appearances – it was natural for the mothers to devote themselves to their children while they were young.

In general I have to say that the position and attitude of women is actually one of my most powerful impressions from the second visit. It is as if their role has changed from the person who creates a supportive home environment to the person who holds up and reaffirms the family's way of life. This was particularly obvious to me in three families where the men have become discouraged ten years on,[223] with one of them even contemplating whether it might not be better to move back to the city. "My husband had health problems and for a year now he's been travelling around America with a friend at the invitation of some people he knows. He works for a bit, travels for a bit – he's finding himself."

In the interviews with the Colourful women, there wasn't a hint of anything I would describe as resignation or disillusionment ten years on or waking from a naïve dream and dismissing the foolishness of one's youth. None of my female respondents expressed the intention of giving up living in these difficult conditions and moving back to the city. Their satisfaction with their life and calm reflection on it, even its more difficult aspects, hadn't changed.[224] When I think of one widow and one divorced woman who were selflessly raising three daughters just on their own salary and meagre welfare benefits, with obviously successful results, I have no qualms about saying that ten years on the women have become the heroes of the research.

Although the women now have jobs, not all of them head out to work in the morning. Some of them stay at home but don't spend their time cooking and cleaning. Two of the mothers work as their children's teachers in a home schooling programme. I was at the Jandas' on the 2nd of September and it was interesting to see what the beginning of the school year was like. The children's teacher/mother made some buns with her pupils/daughters. (Very good, as we were able to ascertain for ourselves.) To keep the inspectors happy and ensure the curriculum was covered, there was physical education (a series of exercises) in the meadow behind the house, and then as part of home economics the dough was kneaded and baked.

One of the objective ways in which the lives of the Colourful have changed over the years is the fact that in some of the families there are older parents who are in poor health. For Mrs Jitka Stehlíková, for instance, this is something which, along with home schooling, determines the course of her days to a large extent. But it is also a valuable experience for her mother's granddaughters. When they grow up, retirement homes and other shelters for 'the aged', which are a common part of Western society, will remain strange and difficult-to-accept institutions for them.

Free of financial worries

How has the economic side of the Colourful households developed? Ten years ago, many of them had to make do with an income verging on a subsistence level. The Colourful had often given up a career

↑ In the second stage of the research, I saw the women differently: no longer just in the traditional role of mother and housewife (Dana Plojharová).

→ Ten years on, I described the women as the heroes of the research (Helena Jandová).

[224] This also applies to most of the Colourful men.

which they had studied for but which didn't fulfil them. They had found work which was meaningful for them but barely put food on the table. In 1992 I spoke to graduates who had set aside their degrees to take up a craft, forestry work or loss-making private farming. I spoke to women who had left their jobs to fully devote themselves to their children and home without worrying that "it's impossible to get by without a second salary". All of this formed the voluntary basis for their modesty.

These days there are major differences between the Colourful. However, they are no longer on the breadline; they are "putting a bit aside" – some just a little, others more.[225] Still, they can all afford to give their children a university education and gradually do up smaller or larger properties. The Plojhar family have finished building the large house that was under construction, and the Nový family are building a new house on the slope above the river valley according to a sophisticated environmental project with the help of their parents.

How was this possible in a society where it is said that the rich get richer and the poor get poorer? If that were really the case, the Colourful would now be living from hand to mouth. One factor which certainly contributed to this change for the better was the fact that the households had gained one more salary thanks to the women going out to work. However, there was also a slight improvement in the financial position of families which had lost a wage earner due to the departure of a male partner (through death or divorce).

It is interesting to note that – with one exception – the existence of the Colourful is not dependent on grant support, as paradoxically tends to be the case with some radically alternative ways of life. The improvement in the families' financial situation was certainly not due to the Colourful having gone back to the professions they had given up ten years ago.[226] On the contrary, they have persevered with the work they turned to as a hobby at the time. They are not performance-driven people, but they are creative and capable. They have begun to thrive as freelancers and entrepreneurs in open and variable conditions of existence. Over time a demand has built up for their products.

They have made a success of something they enjoyed and still enjoy: crafts, artwork, book publishing. Václav Franče has established himself as a producer of decorative leather objects, and Mrs Iva Kozáková sews linen tablecloths with bobbin lace appliqué. Ivo Stehlík is an author, publisher and distributor of books. Jan Svoboda's career path is particularly remarkable. After giving up his graduate profession, in

[225] One family had undergone a particularly drastic change, actually finding itself in a state of deprivation a number of years ago, as described by one female respondent in keeping with my distinction between deprivation and poverty (Librová 2003, pp. 29-30). This unfortunate situation was later resolved through the inheritance of a fairly large sum of money.

[226] Only one respondent was an exception to this. After leaving his job in a pharmacy, he spent a short time working in the management of a landscape park "from idealistic motivations" but decided to go back to his original job due to a lack of money – at least, he says, until his children grow up. Then he would like to devote himself to painting and working with wood and stone.

Václav Franče has established himself as a producer of decorative leather objects.

1992 he began to make his living carving nativity scenes and painting for pleasure. He is now a successful artist and co-owner of an art gallery.[OK]

It is no coincidence that the Colourful's need for meaningful work has been met by employment linked with nature conservation. In 1992 Martin Janda was working at the council offices in Volary and now he is an ecologist/entrepreneur. He is heavily involved in revitalizing excavated peat bogs, which used to be of considerable hydrological and ecological significance for the South Bohemian landscape. Ivan Zwach, who gave up private farming – which offers poor prospects in today's conditions – elevated and expanded his lifelong hobby, an interest in amphibians, into a profession. He helps to deliver projects that aim to reintroduce water into the landscape.

In other cases the Colourful were able to use the proceeds from their business to maintain property restituted to them. The farmer Daniel Maláč even expanded his original six hectares to twenty-five, and his group of six head of cattle has now increased to forty.[227] The Župka and Svoboda families own and cultivate a patch of forest.

A major reason for their tolerable or even good financial situation is the fact that the Colourful know how to make do with less; they know how to budget, as I described in detail in *The Colourful and the Green*. They cook from scratch instead of buying ready meals; they grow their own fruit and vegetables; they make jams, marmalades and compotes; they alter and repair their clothes; they know how to repair simple household appliances and they have a fondness for second-hand

[227] However, his financial situation is not good. For example, there are problems with the local dairy which had been buying up Maláč's milk.

↑ In 1992 Martin Janda was working at the council offices in Volary, and now he is an ecologist/ entrepreneur who is involved in revitalizing excavated peat bogs.

→ Daniel Maláč expanded his original six hectares to twenty-five, and his group of six head of cattle has now increased to forty.

228 Inevitably increasing their ecological footprint in the process.

things. They are even able to make furniture.[OK] At the same time, though, I have to say that this aspect of their economic activities was much more marked ten years ago.

Both male and female respondents admit that since the women have gone back to work and their arts or crafts products have taken off, they are no longer such 'models' of environmentally friendly frugality and modesty. For example, they have cut back on growing vegetables and breeding animals and making full use of the fruit harvest. They go to the shop more often. These days the successful craftsmen, artists or teachers primarily devote themselves to their work, which they enjoy and make a decent living from. They no longer maintain a plot of potatoes big enough to supply their needs. Working in the garden has become more of a hobby for them – perhaps even an ecological luxury.

No escaping the car and computers. A well-equipped kitchen

How have the fixtures and fittings of the Colourful households changed? On arriving for the field trip, I found myself in familiar spaces. In 1992 they hadn't been settled in yet, because most of the Colourful had only recently moved. Over the space of ten years, the house or flat has been converted with great effort and considerable investment of materials and energy[228] and furnished to suit their taste. Many things have been newly installed, particularly the functional heart of the home: the kitchen. However, the basis of the flat still revolves around

169

the original furniture and smaller items, which have been inherited, found or given to them, as well as objects purchased from junk shops. This is not just a stopgap measure. The Colourful have retained a fondness for old things, for objects which remind them of the person who gave it to them or the circumstances in which they were acquired. "Both of us are scavengers," says Roman Kozák. "My wife is a hoarder. Nothing gets thrown out in our house. We can dress up as whatever we like."[229] "It's all second-hand, brought down from attics, given to us or found." Many of the families have acquired more musical instruments as their talented children have grown.

The car began to figure more frequently. In 1992 about half of the Colourful had one. Ten years on, with one exception, all of them have a car – even those who had spoken about cars with contempt in 1992. However, the Plojhars and others stress that they don't use the car all the time, but only when necessary. Mr Plojhar travels to work by bus, and his children and his wife Dana (even though she's a businesswoman) walk. As well as the reason commonly given – a lack of public transport and living in a remote location – there was another decisive factor for some of the Colourful: they needed a car to take their arts and crafts products to the market.

The family of the artist Mr Svoboda actually has three vehicles: a big one for transporting paintings and a small one used by the grandfather, wife and son. The third vehicle is an "American classic car", the "great America" which Jan Svoboda has as a personal "whim" of his and which continually needs money invested in it for repairs. However, it also serves some kind of practical purpose: the Svoboda family use it to transport wood from their forest to their own sawmill.

It's not only the car that is important for the Colourful living in remote places. A key existential issue for them is decent road access: a road or at least a track that links them with the rest of the world – something which might not even occur to a city dweller. But after having made the trip along those bumpy roads in our research car, it didn't surprise us that time and time again our conversations with the Župka, Vaněk or Plojhar families turned to the topic of 'the journey home' – not only in the abstract sense, but more importantly in the physical sense.

The bicycle is a permanent feature of the adults' and children's lives, just as it was ten years ago. The train dispatcher Roman Kozák says: "The bike's a part of me, like the train. I ride my bike every day. I couldn't live without a bike."

229 This is meant literally there are theatrical elements to the Kozák family's social and business activities. When they sell their wares at craft fairs, it is not just the quality of the goods that attracts people to their stall, but also the stallholders' get-up – one time as a couple of villagers from Chodenland, another time as a Jewish merchant. Roman Kozák also dressed as a dandy from the early twentieth century to race on an old bicycle with a locomotive.

Over the past ten years, my guess that the Colourful would develop a penchant for modern technology was confirmed.[230] Most of them had got themselves a mobile phone, something which is invaluable given the remoteness of their homes, the need to stay in touch with family and friends and their involvement in organizing public life. The Colourful like working with computers and some of them have email addresses and look up information on the internet.

A deep freeze is essential for the running of most of the Colourful households. It allows the food produced by the Colourful in season to be efficiently disposed of year-round. In the context of a book about ecological luxury, it is significant that those among the Colourful who are in a good financial position have gradually equipped their household with environmentally friendly technologies. A separate chapter – and a characteristic one for the life and attitudes of the Colourful – is the dishwasher. I shall come back to it later.

On the subject of housekeeping, it should be noted that in 2002 all of the Colourful separated their waste. It would be unthinkable for them to throw a mixture of everything into the bin. I was surprised to find that waste management operates better in villages and small towns than it does in the city of Brno. Often it was my respondents who had pushed for this in their communities.

Lia Ryšavá strives for self-sufficiency in the household.

Do children spell the end of a modest life? A dream holiday

Sceptics claim that life in a small village, never mind in a secluded location, ceases to be tolerable at the point when the children begin to grow up; when they start going to primary or secondary school, when they want to attend after-school clubs. Supposedly, poor bus transport links will force the Colourful to give up life in a small community, whether they like it or not, and move to a larger village – or, more likely, to a town.

Here too the Colourful have evolved in various ways. As we shall see shortly, school attendance did not change the Vaněk family's remote place of residence. The experience of the Plojhar family, who also lived in an isolated location, was different: It was far from ideal that the children had to travel alone by train to České Budějovice to get to their after-school activities, returning home late in the evening after dark. Although they did sometimes wonder whether the children might not be better off climbing trees, in the end they did move.[231]

Another concern is that children living in a remote area are deprived of contact with their peers. This might make them feel lonely if they don't have siblings, which isn't usually the case with the Colourful. Children from remote areas seem to command a special kind of respect, or perhaps romantic admiration, from their peers. What I saw on the outskirts of Volary reveals a lot about the life of the Colourful children: Despite the fact that the children live in semi-solitude and are taught at home and don't have contact with classmates, the Stehlíks' house is simply swarming with friends from the immediate and wider surroundings.

Another question comes up in discussions about the Colourful. It relates to the status of the children within their peer group: "How

[231] Of course, as we already know, proximity to a school was not the only reason for their decision.

Will the erosion of voluntary modesty begin with the discomfort the children feel in school in the competition over fashionable clothes? (Marie Bukáčková-Matějková with her daughters Marie and Františka; Anežka is not present.)

172

do (or will) their children feel about this?" Meaning: what view will the children take of their parents' modest lifestyle as they grow up? How will the boys and girls dressed in non-designer clothes react to their classmates' competitive attitudes? Isn't it possible that the erosion of voluntary modesty in the families will begin with the children, with the discomfort they feel in the classroom setting? Will the parents give in to it?

When I steered the conversation towards this topic, the Colourful responded in various ways. Some of the children really did find it tough when their classmates talked about television series they hadn't seen or flaunted designer jackets their modest parents were unwilling to buy them. This particularly applies to younger children attending primary school. Children with a number of siblings are more resistant to it, especially if they have older brothers and sisters. Colourful parents do not respond in a principled, hard-line way, but 'negotiate their way' around the situation. Now and then they relent and buy children the clothes. Mostly, however, they rely on the children's attitudes being positively affected by age. That is also why they advised me "you ought to come back in another ten years".[232]

However, according to most of the Colourful parents, their children do not have problems with a lack of fashionable clothing and equipment. The popular view which emphasizes the irresistible force of peer pressure may be exaggerated: indeed, I know of children who manage to impress those around them with their pointedly non-consumerist behaviour. I can easily imagine the confident children of the Colourful being capable of such non-conformism, or even inspiring others.

Among the influences that have altered the lifestyle in post-communist countries in the last decade, perhaps the most conspicuous is the organizational and financial accessibility of foreign travel and ubiquitous advertising for a 'dream holiday'. What impact did this most powerful of consumerist enticements have on my respondents, most of whom had claimed not to be tempted by foreign holidays in 1992? The Colourful did not automatically dismiss the opportunity to travel and decided to try it. However, they did not use the services of travel agencies and did not go far – most often to Croatia.

To a greater or lesser extent, they describe the travel experiment as a disappointment. Mr Václav Franče, whom we got to know as a 'Euro-Indian' in 1992, recounted his disenchantment with a trip to Italy in typically sweeping terms: "I said to myself, we'll find some kind of forest there so we can sleep under the stars, and after three days we discovered that for one thing we couldn't find any kind of forest, and for another all the land was privately owned – they had those gates that close in front of you... The brochures made it look as if Abruzzi was 4 km from the sea – there are all these pictures of forests with oaks and plane trees, so you think it's just around the corner, like it must be within walking distance. But it's actually sixty kilometres inland – the brochures are lying." In a rather non-Indian way, Mr Franče was looking forward to seeing the historical sights:[233] "So we reckoned we'd arrange some kind of minibus and the whole lot of us would go and have a look at the cathedral in Atri, visit Teramo where the famous pottery comes from, finally take a look at that park in Gran Sasso where Mussolini was imprisoned, see some of the sites connected with the Romans – that we'd really get to enjoy a good chunk of Italy. I was so naïve that I thought we'd see Perugia, go to Assisi and Lake Trasimeno... It turned out that the only thing you could do was walk along the beach and collect pinecones and shells."

Mrs Jasmína Vaňková hasn't been disappointed with her holidays. For more than ten years, she and her family have been going on canoe trips every summer, but in recent years she's had two holidays in Croatia, where she spent several days sailing between the Kornati Islands with several friends. It was glorious. In 1992 Jan and Magda Svoboda stated that a foreign holiday didn't appeal to them at all, that they only travel when they have to, but now they say: "We've discovered the sea." They go to Croatia in the off-season and are usually met with the autumnal aspect of the sea, but they like to take advantage of it. Jan always finds subjects to paint there.

The Colourful don't go to foreign destinations further afield, although they sometimes send children to relatives there. This is another example of how they try to provide their children with the best possible education and broaden their horizons.

Disappointment with public and church life

The past ten years have been marked by a general disillusionment with the way society has developed and a decline in the interest in civic life which arose in the late eighties. This development has not bypassed the civically engaged Colourful either. Most of those who were elected to local councils in the first elections after the Velvet Revolution did not stand for re-election. However, they did concede that they would give it another go in the future if the political situation in

[233] I admired the way Mr Franče reeled off the names of Italian monuments with aplomb, accompanying them with a cultural-historical commentary, even though he hadn't been able to see them for himself and two years had already elapsed since his trip to Italy.

Magda and Jan Svoboda are in search of a spiritual shepherd.

When she wasn't mayor, Mrs Válová dedicated herself to the development of the Konstantinovy Lázně microregion. However, she was the mayor for most of the terms of office.

[234] This strategy underlies the career choices of the Colourful. In a similar way, they had taken the decision to leave their professions which they were unable to identify with. They are unwilling to make pointless compromises, so they go off in directions that seem meaningful to them.

[235] It's interesting how these words mirror the statement made in 1992: "My reservations about the church as an institution don't go so far as to stop me going to church, because what I get out of it is stronger than my reservations."

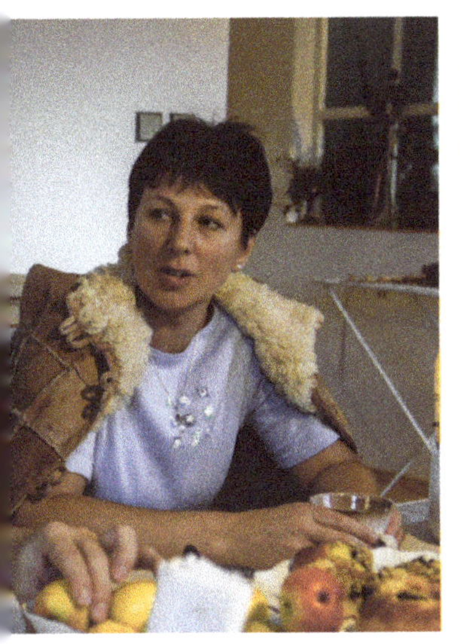

town halls were to change for the better. Perhaps in four years, at the next elections.

However, leaving the official structures of elected office did not mean that the Colourful were withdrawing from public life. They play an active role in their communities and cultural life in other ways which do not involve their efforts being ground down by the ineffectual voting of local councils and the backroom wrangling of party politics.[234] They lead Scout troops, run art clubs and carving courses, organize town festivities and bring out local newsletters. It could be said that they have returned civic society. "I'm at the town hall more often than the councillors. I work alongside the cultural department and it seems to me that I'm more use to Volary than I would be sitting in council meetings, discussing bylaws and the sale of plots of land," says Roman Kozák. When she wasn't mayor, Mrs Válová dedicated herself to the development of the Konstantinovy Lázně microregion. She was an advisor to eight small municipalities. For her, this was not just about looking after the built-up areas of municipalities, as it had been up to then, but also the countryside that surrounds them.

It is probably no coincidence that some of the Colourful no longer read the daily press. They say they are tired of information that is only meaningful for a day at most. Moreover, they are irritated by the way the newspapers keep growing in size.

I observed one other sign of disappointment in some of the Colourful: disappointment with religious life bound up with the church. Because many of the Colourful are deeply religious people, they react to it with sorrow, and at other times with a certain sense of liberation. Jan Svoboda: "These days the things that don't sit well with me outweigh the positive ones.[235] So basically I don't feel the need for contact with a spiritual movement like Catholicism. So I've called it quits. I can do without it in my life It's not my style just to play along with it as if everything's OK." His wife Magda says: "It's also to do with the fact that the parish priest we used to go to in Fryšava retired, and since then there have been a few different priests there. And because we'd been going there for a really long time and he was a really good preacher, he set the bar awfully high. And there's probably no point going to the service when you know you're not going to get anything out of it, that the sermons you hear are just empty phrases and when you look at the people around you, you can see that they're only there because they're used to going to church every Sunday. So for a long time we flitted from one place to another – we went to Nové Město, to Žďár, and about three years ago we discovered a clever young parish priest in Moravany, but six months later he was posted to the Vatican, so we were left with nothing again. And since then, when we go to church, we're actually just going there for our own, personal kind of spiritual feeling and we don't stay long. When we do catch a sermon, it makes us sad and we say to ourselves: 'Well, we knew what we were in for.' So it's a shame."

Six snapshots of luxury

In 1992 the Stehlík family were struggling to make ends meet. These days it's better. In 1994 Ivo Stehlík left the post of deputy mayor of Volary and for some time he made a living working with a chainsaw in the woods.[236] At present he runs a business as a publisher and distributor

Ivo Stehlík with the rescued mare.

236 He learned about the issue of felling in the Šumava National Park at close quarters and formed a critical view of it.

of books. He himself has written two books inspired by the setting of the Šumava mountains.

Mr and Mrs Stehlík have an affinity with ecological initiatives; their household set-up is built on environmentally informed rational thinking.[UK] For example, they don't have a refrigerator, an item owned by practically every Czech household. They decided that they didn't need one since they have a cool corridor. On the other hand, a deep freeze is indispensable given their semi-agricultural operations.

The Stehlíks have work to do every day on their small farm, especially with the animals. Thanks to their efforts, the household is largely self-sufficient, and completely so in meat consumption. Mr Stehlík is able to slaughter the animals, although as the years go by he is increasingly reluctant to do so. However, I also detect a certain pride in his words, a sense of superiority towards those of us who eat meat but avert our eyes from the fate of animals and the horrible end they meet in the slaughterhouse. So the subject of abattoirs came up:

"We have a small abattoir a hundred metres behind the house. One time I was on my way back home and I noticed there was a horse walking around the abattoir site on the other side of the fence. It never crossed our minds that the mare was there to be slaughtered. You don't make the connection – I thought that's brilliant, one of the butchers is learning to ride a horse, that's great. But the mare had been put there as a two-year-old because she was blind, to be slaughtered. The slaughterhouse people didn't want to kill her. They could have done it when they got her out of the van, but they didn't want to kill her. They were trying to find a new home for her – they wouldn't have done that for a pig or a cow... The people there are affected by their job – they can't help being – although not completely. So we ended up with a beautiful horse. Our vet cured her in ten days. Unbelievable! Even the experts don't want to believe it, because she had what's called moon blindness, which vets consider incurable. He did it using homeopathic medicine, antibiotics and special care. And we ended up with two horses: she was in foal."

Now the mare is grazing in the meadow behind the Stehlíks' garden in the company of six other horses. Some of them are used for riding and two of them for working, but the others live just as they are, for the sheer joy of living, for the joy they bring to the Stehlíks and their friends[237] and as a beautiful component of the sloping meadows of the Šumava region.

Mrs Jitka Stehlíková teaches her two daughters in home school. At the Stehlíks' this domestic institution has one extra special feature:

they have a young woman living with them who left Prague at the age of eighteen with her young daughter. She earns money giving private English lessons to adults and also teaches the Stehlík children English in return for room and board. Even with all the work around the house and the farm, don't the Stehlíks live in a style of luxury to rival that of the Buquoys?[238]

***

The Vaněks live in a tiny hamlet called Kořen. These days their two sons are studying at the grammar school in Tachov and their daughter at the Evangelical Academy in Prague. When the children were younger, the journey to school wasn't an easy one. They had to walk along a rough forest track to Olbramov, three kilometres away, where they joined their classmates to take the bus to the primary school in Černošín. When they were still in first and second year, they were accompanied by one of their parents – either on foot or in a horse-drawn cart.[239] However, as soon as the children got a bit older, the three of them began walking to Olbramov by themselves. "They'd pop in to say good morning to me while I was still in bed, I'd say good morning to them, and then they'd leave. It's about half an hour through the woods, forty-five minutes in the winter. When the snow was deep, the kids had to forge a way through... They're really good that way, I have to say... They'd set off before half past six, because the bus left at seven.

They had it all precisely worked out, what time they had to be where – they had to get to the pine tree by such-and-such a time, get to Davídek by such-and-such a time, but they knew how to do it so they were on time for the bus. And if they did miss it, they'd keep going, and they'd actually learned to walk so fast that when I occasionally had to go with them, they were way, way ahead of me and I was panting along behind them... When the older children grew up, Ondra walked to school by himself for a year. By then he was in sixth or seventh year. He never complained – at least, not out loud... They'd been walking to school since first year, so they'd never known anything different... They always got there – even when the bus didn't run in winter. The kids from Olbramov wouldn't arrive but our children just kept on walking till they got there... When the kids reminiscence from time to time and talk about the past, we find out how they used to play along the way – they had various games. Like when Míla took them in the cart, they'd play a game that here in Kořen was the past, somewhere

Jasmína Vaňková graduated in special education and teaches at a special school in Stříbro.

[238] The Buquoys were a powerful aristocratic family of French origin who acquired property in Bohemia in the post-Thirty Years' War period. Among their ranks were men with philosophical interests, but also excellent state economists and shapers of the South Bohemian landscape. Count Georg Franz August von Buquoy established a reserve that became the first of its kind on the European mainland – what is now the Žofínský Prales nature reserve in the Gratzen Mountains.

[239] As shown in the photograph on page 146.

178

along the way was the present and somewhere up ahead would be the future… It used to take them a lot longer to get home… They had to be home by five so I wouldn't start to worry where they'd got to. The bus got into Olbramov at three, so they had two hours to get back. They'd stop by the stream, lie on the hill and sunbathe. I think they enjoyed themselves. In winter, when they went down that hill, it was like a sheet of ice. So they'd walk with a stick and count how many times they fell."

I think about the luxurious experiences the Vaněk children had on their long trek to school every time I see the children in our neighbourhood getting out of cars and stepping straight from the door of their luxury cars onto the steps by the school door.

***

Ten years ago Mr Kozák was described by a colleague from the town council as an ascetic. Even then he protested; he didn't see himself that way. These days he says that his second wife got the measure of him, that he is actually a hedonist.

"We make up something good and then we put it into practice." Roman Kozák is the author of several books about railways and about the local area. He was involved with an amateur theatre troupe in Volary as a playwright, director and actor. Now he is an initiator of cultural

"The advantage of a small town is that whatever we come up with can be put into practice. Once a month we take our wares to local fairs, all decked out in their finery – why would we trek halfway around the world when we have everything right here?!" (Iva Kozáková and Roman Kozák).

life. He creates hiking trails around the town and organizes bicycle races with a steam locomotive. "We've got this association with friends where once a month we stage amazing events that we look forward to for six months beforehand and then we document it and put on exhibitions."

Apparently, the advantage of a small town is that "whatever we come up with" can be put into practice. And also that it offers plenty of time for life – "If I want to go for a walk and bring home some mushrooms, I only need an hour for it. I can get to work in a couple of minutes by bike." This is what Mr Kozák told me about his holiday in Greece: "It's not all it's cracked up to be – the heat almost killed us and there isn't a decent bit of woodland." And he goes on: "Years ago I got it into my head that I'd like to cycle round Scotland. But then we discovered Scotland in the Šumava region. Low stone walls, heather everywhere... When our friends ask where we're going on holiday, we say our life's one long holiday. When the weather's nice, we go to Lipno Reservoir. When it's like today, we ride a bike in the woods. When it's raining, we go mushroom-picking. Once a month we take our wares to local fairs and we see castles, towns and squares all decked out in their finery – why would we trek halfway around the world when we have every-thing right here?!"

***

In the autumn of 2002, the Jandas' family discussions were centred around the question: To buy or not to buy a dishwasher? Although the Jandas are among those from the Colourful group whose palette is dominated by the colour green, they do not share the radical view that regards this appliance as a symbol of consumerism and extrava-gance and an unnecessary indulgence – there's no need to bring out, or even to own, that many dishes. Mrs Helena Jandová says: "I've been giving it a bit of thought. The reason being that we're starting to get more visitors.[240] And I think it would save me time, which means I'd have longer to spend with the guests. But I'm not sure, I'm still making up my mind." Mr Martin Janda adds: "If we were to buy a dishwasher, we'd probably get the best one that's around just now. Something that ticks all the boxes – that means low consumption of both electricity and water. Then it's comparable with normal washing.[241] But it's some-thing we still have to consider very seriously."[MK] And his wife says: "I don't like how you have to put that rinse aid stuff in it. I don't know

[240] The Jandas only recently moved from a block of flats to a country farmhouse. In this respect they differ from most of the Colourful, who went through a similar stage of life more than ten years ago.

[241] This is compared to washing dishes in a sink and rinsing them. Compared to washing them under running wa-ter, the dishwasher is even better.

"I just put this together on a kind of whim, so I can walk past the boiler and see the water getting heated up. But we also have to get the wood ready, obviously." (Jan Svoboda)

if a dishwasher would work without it. It probably would be a really big help, because there's more work to do here. We want to spend as much time as possible outside. I don't know, I'm really not sure if we'll go for it in the end."

In the extract from the interview, there are two aspects worth noting: the significance attached to visitors, and the long and responsible hesitation over purchases, which contrasts with 'I've got to have it' impulse buys.[242] In some of the other families, the dishwasher is considered an excellent thing and praised in connection with large numbers of visitors. For example, the Svobodas particularly appreciate it on Saturdays and Sundays, when the extended family and friends get together – sometimes as many as fourteen people.

***

The Svobodas and Župkas have a soft spot for eco-friendly technologies. Jan Svoboda has a keen interest in advances in environmental technology both as a trained engineer and as an artist. He bought a boiler for burning wood and wood waste efficiently. "We've got a new wood-fired boiler and I had this huge storage tank made, so we actually heat up a five-cubic-metre barrel. So today we're burning wood for heating and then for two or three days we use the heat from the barrel. For one thing it's big and for another it's well insulated so the heat doesn't escape, and those five cubic metres of water get heated up to about eighty or ninety degrees, which keeps those two flats of ours warm. In winter we might use the heating every other day, and in the in-between seasons every third day.

Here in Radňovice the village got connected up to the gas when I'd spent about two hundred grand on this thing that I'll never get back. And so everybody told me, you know, with gas you just shell out seventy grand for a boiler and your worries are over – all you have to do is press a button. So they press their button and I've shelled out two hundred grand for this welded-together thing and the boiler and so on, and now I'm happy with it. We want to use wood for heating, that's for sure, because we've got a forest, so what else would we do with all that wood. I enjoy it – I enjoy working with wood. So I just put this together on a kind of whim, so I can walk past the boiler and see the water getting heated up. But we also have to get the wood ready, obviously."

Mr Svoboda devised, designed and began to draw up plans for a wind turbine, "a latticed thing like something out of Jules Verne",[243]

which was supposed to cost a quarter of a million. In the end, the plan
was dropped because a more detailed study of the natural conditions
revealed that the winds in the Bohemian-Moravian Highlands are
irregular.

Now Jan Svoboda has developed a keen interest in photovoltaic
technologies. He regards them as promising, is following their development in Germany via the internet and would be willing to use them
in his house if they were at a more advanced stage of development.
"I'm really excited about it. Here they're building those stupid Te-
melín-style nuclear power stations. It bothers me. It bothers me from
the simple point of view of economics and physics, because it's an
utterly pointless waste of resources, an inefficient way of producing
energy. With a different political establishment – but I can't imagine
that anywhere in the world – the money could be used to obtain the
energy in a much nicer way – I mean from renewable sources."

***

The sixth example of a modest luxurious life: Lia Ryšavá was widowed
in 1994. For years the family lived off her teacher's salary and the children's orphans' allowance. Two of her daughters went to university
and are now married. Michaela, who is disabled, remained at home.
When we visited the Ryšavý family, preparations were underway for
a public concert in the Protestant church. "It's Michaela's birthday on

Tomáš Ryšavý died in November 1994. As the foreword to this book testifies, he is still missed by us researchers too (in the photograph his wife Lia and daughter Michaela).

244 The Colourful's way of life fully affirms the significance of temporalities – activities and things clearly and consciously associated with the passage of time – in ecological luxury as well as the words of Ulrich Beck: "[C]ontrol over a person's 'own time' is valued higher than more income and more career success, because time is the key that opens the door to the treasures promised by the age of self-determined life: dialogue, friendship, being on one's own, compassion, fun and so on" (Beck and Beck-Gernsheim 2002, p. 161).

The Colourful's concept of time in 1992 was characterized by the blurring of the line between work and leisure time. Where the man is working as an artist or craftsman, this classic feature of all alternative visions of a good life has been preserved even today. Where the women have gone out to work, however, it has been disrupted. Some of the mothers have managed to partially preserve it by remaining at home as home-school teachers.

Sunday. Michaela likes Slávek Klecandr, the lead singer from the band Oboroh. Oboroh have been in Miroslav a few times, and Mr Klecandr is very understanding when it comes to these things, so I was thinking about how to make Michaela's birthday as happy as possible, because she has different ideas about happiness in life than we do. It wouldn't make her happy if she got a new dress, for example. It's important to choose presents with her interests in mind, and her interest is music. So I was thinking about how to do it, and then I phoned Mr Klecandr to ask if he was free on the afternoon of 17th November and if he would accept an invitation to put on a concert in Miroslav. And he was free that day. Obviously, I called him very early on, some time in the summer. We're organizing the concert ourselves and obviously the general public are invited to it. If this is a luxury, then we live very luxuriously."

The Colourful 2002 – ecological luxury – voluntary modesty

It was interesting how the Colourful responded to my direct question: "Do you think your life contains anything that you would say was a luxury?" Some of them responded immediately, while others were momentarily taken aback and took issue with the word luxury. Once we understood each other, they knew immediately:
- "our luxury is having enough time"[244]
- "here you're outdoors as long as there's enough light to see"
- "you might be busier here than in Prague, but you decide for yourself when to do what"
- "I don't have to drive anywhere to enjoy nature – I have water, forest and meadows here"
- "a sense of privacy, peace and quiet"
- "we have clean water and healthy food from the garden"
- "in the countryside work recharges my batteries; in the city it's the opposite"
- "the greatest luxury is being able to do something I enjoy, that makes sense to me"
- "we don't have to worry about the children playing outside"
- "the feeling that I'm not actually missing anything here at all"

The answers given by my respondents were in line with the categories formulated as luxuries at a theoretical level by Hans Magnus Enzensberger (1996): today's scarce commodities are time, attention, space, peace and quiet, nature and security.

The Colourful's attitude is characterized by the self-restraint of ecological luxury: 'you can't have and experience everything'. They have achieved what predatory luxury is incapable of: the ability to say NO to opportunities. Thanks to this, they have gained time, attention and the ability to marvel, space, peace and quiet, nature and security. As we already know, many of them resigned from a steady job and a career more than ten years ago. They gave up the benefits of city living: trouble-free supplies, proximity to schools and easy access to medical care. They left behind a comfortable city flat and moved into a dilapidated rural building which they are laboriously fixing up with their own hands. Perhaps the most distinctive feature of the Colourful's way of life is their friendliness towards other people, their civic-minded and more generally humane sense of responsibility.[UK]

Of course, there remains the question of whether the Colourful fulfil the second condition of ecological luxury: the deliberate reduction of their ecological footprint. Yes, to a greater or lesser degree, they try to do so, although they wouldn't describe it in those terms themselves.[245] In contrast to the situation ten years ago, the Colourful reflect more on the environmental aspect of their lives.[246] They try to reduce their ecological footprint through the structure of their consumption patterns and by favouring biophilic household technologies.[MK] In their gardens and in the wider countryside, they manage things in a way that is sensitive to nature. Many of them have become involved in greening

[245] When the talk turns to environmental issues, some of the Colourful take on a defiant look. It seems to me that prescriptive green ideology does not sit well with these unconventional respondents with their firmly composed attitudes and that in the interviews they take up an adversarial position towards what they assume my environmental agenda to be. In other sociological studies the opposite is true: the respondents try to comply with the presumed expectations of the researcher. When interviewing the Colourful, I came across a methodological rarity, a kind of self-stylization turned upside down.

Jan Svoboda remarked: "When Rosťa found out you were coming here, he advised me to hide the fridge!" However, this was too much for his wife: "This environmental awareness is part of him too. When we go to the shop and the shop assistant starts putting his things in a plastic bag, he says: 'No no, don't give me a plastic bag, I don't want it!'"

[246] Without wishing to flatter myself, it is possible that they picked up a few things from reading *The Colourful and the Green*, in which they were the protagonists of the book, or perhaps from the subsequent interest journalists took in their way of life.

[247] To be found in *The Half-Hearted and the Hesitant* on pp. 154-162.

public spaces; for example, they have helped to promote the sorting of waste. Some of them have chosen jobs that are directly related to nature conservation.

Let us go back to the initial question and ask: Have the Colourful conformed to the society around them? Would they still be a model of voluntary modesty today? The answer may emerge from the preceding pages. All the same, I would like to offer a summary:

I found families whose resistance to the erosive forces of the society around them has endured or even increased. Mr Daniel Maláč, a young farmer, continues to live modestly; as far as possible, he persists with his work, which is regarded by his carefree peers – not just the urban ones, but the rural ones too – as incomprehensible drudgery. He breeds his animals with unwavering faithfulness,[UK] inner joy and the greatest possible respect for nature. The Janda family moved from a flat in a small-town housing estate to a farmstead on the outskirts of the town and radicalized their way of life with a move towards self-sufficiency. The father left a clerical job in the council's environmental department and is now involved in ecological field work. The Stehlíks have also held firm; as we saw in the first snapshot, they are capable of balancing self-restraint with high expectations of life in a way that is clever and respects the environment – a skill that we view as an ecological luxury. The family of Mrs Ryšavá, with its grounding in the Christian faith, has not made concessions to consumer society in terms of its lifestyle either.

Nevertheless, consumption has increased in most of the Colourful households. This is not just due to the simple fact that older children cost the household budget more than toddlers, that benevolent parents give in to children when they come home from school dejected because they couldn't compete with the others' fashionable clothes. The ecological footprint of households is also increased by other factors: for example, the reduced use of public transport which goes along with the purchase of a car. The indirect consumption of energy and materials has grown as there has been less intensive work gardening, making preserves and mending things since the father has immersed himself in his work and the woman has taken up employment.

Modesty and environmental virtues are threatened if the household comes into money. The case of the 'Dutch doctors' (Aarts 1999) shows[247] that even where reducing consumption and the ecological footprint is a plan, or indeed a way of life, it is not easy to manage money in an environmentally friendly way.

The Colourful are people with strong inner integrity and an unusual depth of inner world. Consumerism goes against the grain for them. Yet some of them have abandoned the level of modesty they were voluntarily living in ten years ago.[248] How could we expect millions or even billions of more earth-bound individuals to permanently opt for voluntary modesty as their way of life, as envisaged by some concepts of sustainable living?

Even so, the Colourful are significant for reflections about environmentally friendly living. Through their way of life they illustrate that Bauman's metaphor about moving sands does not have universal validity. Of course, modern society exerts pressure on us that standardizes our lifestyles; despite all the diversity of goods and holidays, the lives of consumers are monotone and similar to each other. The Colourful's greatest luxury is that they don't allow themselves to be dragged along and manipulated; they act freely. They respond in different ways to economic, social and political pressures based on their personal inclinations and the interpersonal configuration of the family. As far as possible, they shape their lives themselves, on the basis of a well-intentioned and thought-out attitude.

This corresponds to two characteristics which Ulrich Beck expresses as 'do-it-yourself biography' and 'reflexive biography' (Beck and Beck–Gernsheim 2002, p. 3). I hope he isn't wrong in attaching great importance to them: "The ethic of individual self-fulfilment and achievement is the most powerful current in modern Western society. Choosing, deciding, shaping individuals who aspire to be the authors of their lives, the creators of their identities, are the central characters of our time" (Beck 1999, p. 9).[MK]

[248] I ask myself whether I would have chosen them for my sample of voluntarily modest households in 1992, and the answer is no.

186

IV.3. The Colourful 23 years on

The third stage of the research: no major changes expected

As reported in the previous two chapters, our research into environmentally friendly life-styles, or voluntary modesty, began in 1992 and continued ten years later. When our hosts showed us to the door in 2002, some of them remarked, "See you in ten years". We took this as a joke, a polite way of showing their appreciation rather than a real possibility.

These exchanges on the doorstep in Volary, Žďár nad Sázavou and Miroslav came to mind in the winter of 2014 when we received the news from the Czech Science Foundation that they would support the continuation of our 'time-lapse', longitudinal (in sociological terms) research in 2015 and 2016.

This time I would not be the sole interviewer and researcher. Conscious of my advancing years and dwindling energy, I would have been hesitant about planning research consisting of in-depth interviews with respondents living in remote locations. Moreover, the logic of the research called for a new dimension: We needed to ask how the Colourful's (mostly grown-up) children were living. The research team therefore expanded to include sociologist Lucie Galčanová Batista, Lukáš Kala (who had recently been awarded a PhD in environmental humanities) and anthropologist Vojtěch Pelikán, who also took on the burden of the complex organization of the project.

The third stage of the research.

During our numerous meetings, we took a cautious approach to revising the content of the interviews with the Colourful twenty-three years on in order to ensure that the results would be comparable over time. We weren't expecting to make any major research discoveries during our visits. The methodology, including the interview format and method of visual recording, had been tried and tested in the previous studies.[249]

On the road our small team was joined by the film-maker David Musil. He knew the respondents from the previous stages of the research and had kept in touch with some of them. David's camera wasn't intrusive, and his initial search for a plug point helped to break the ice before the interviews. Between May and November 2015 we made fifteen visits to homes in South and West Bohemia and North, Central and South Moravia which we had visited in 2002 and which had low economic dynamics.[250] We acquired our information through direct observation and through in-depth interviews which lasted from one-and-a half to two hours.

As in the second stage of the research, we had no problems finding the addresses of our respondents. We weren't surprised to find that our conservative respondents hadn't moved anywhere – and if they had, they had remained in the local area.[251] The respondents' decision not to migrate was the first dimension of their faithfulness,[UK] which we encountered even before setting out on our field trip.

[249] We were able to dedicate the meetings to preparing exciting interviews with the children of the Colourful. The results are presented in Chapter IV.4.

[250] In the first stage of the research, 47 addresses were chosen from the original address list of 113 households so that the research would be feasible in terms of scale and transport. In 1992 there were 44 interviews with 70 people. In the second stage, in 2002, we carried out another typological selection for the in-depth interviews. For further information on the research methodology, see previous chapters.

[251] The Colourful don't necessarily live in the countryside; they are also to be found in towns. We cannot label country life as 'alternative' - on the contrary, it encapsulates the ideal of an Arcadian way of life present in European culture.

Despite the attractive job opportunities in neighbouring Austria and Germany, Roman Kozák remained a train dispatcher in Volary.

Nevertheless, even during this preparatory phase of the research we had the opportunity to discover that the distinctive, conservative Colourful, so faithful to their lifestyle, were in some ways succumbing to the pressures of the time: After that long pause of thirteen years, we deliberately contacted our respondents the old-fashioned way – by letter, not actually handwritten but printed out and sent by post. The response was fast, personal and approachable; everyone replied by email apart from one who replied by mail. The evolution of communication between the researchers and the respondents is interesting: The mainly written and occasional telephone contact in 1992 was replaced almost exclusively by the telephone in 2002 and then by email in 2015.

The constancy of the Colourful was also reflected in their professions. The research from 1992 often recorded radical changes in careers that had occurred after 1989. For example, it was common for them to have left higher education or technical professions and turned to arts and crafts. In terms of Hannah Arendt's categories, the Colourful had gone from work, which is the basis of professional technical activities, to caring for disabled people and the environment, to action in the form of working for the community, and to the contemplation that is part of artistic creation.[OK] But after that the career choice stabilized. The Colourful did not succumb to the professional dynamism of the early 21st century, which was so highly valued and promoted by society. Despite the attractive job opportunities in neighbouring Austria and Germany, Roman Kozák remained a train dispatcher in Volary; Martin Janda continued to protect the environment in a protected landscape area; Marie Bukáčková, now Matějková, was still working as a nurse as she had in 2002 – now a charge nurse in a septic surgery unit; Jasmína Vaňková was still teaching at a primary school in Stříbro. Jan Svoboda was now a prominent artist; Miroslava Válová was mayor once more; and Ivo Stehlík was a book publisher, distributor and author.

If respondents had changed jobs, this was normally due to external circumstances rather than the result of a change in their outlook and a personal decision. After the border with Austria opened, Mrs Karla Frančová (alias Kay) lost her job in haulage. Daniel Maláč, a smallholder running the family farm, withstood the unfavourable economic conditions for a long time before finally succumbing to the competition from large agricultural corporations. Now he was employed at an engineering works in Slavkov u Brna. Vladislav Bukáček's career path had been unusually dynamic. Following his divorce, things had been difficult for him in a small-town Catholic community; he left his job as an engineer at the Žďas company and set up the Kopeček sheltered workshop, which employs 34 people with various disabilities. The career path of Bob Plojhar was preordained: he inherited a cardboard-making business from his father. In 1992, Miloš Švanda was working as an electrician at the Temelín power plant, but since then his physical strength and artistic talent have been directed towards preserving monuments in the Třebíč region. At the time of this research, he was working as a stonemason at the Jewish cemetery.

An empty nest

It has already been said that despite the dynamism of the time, the lives of the Colourful had not changed much over the past thirty years. Our respondents had stayed in their jobs and homes; the basic features of their lifestyle and attitudes had not altered. And yet over the years when the Colourful had dipped under my research radar, important things had been happening: The farmer Daniel Maláč got married! In Volary, the Jandas' daughters Rozálie and Tereza, and the Stehlíks' daughters Barbora and Jolana, were home schooled by their mothers. The father, Ivo Stehlík, became a campaigner for home schooling at a nationwide ministerial level and has written books on the subject. Following the *homo curans* model,[OK] the time spent caring for her elderly parents allowed Mrs Jitka Stehlíková to pass on lessons to her daughters which were just as valuable as the material from the home schooling textbooks. The grandparents' failing health revealed the pain, difficulty, importance and meaning of ageing, as well as the value of sacrifice.

When the grandparents died and the children left home to study, the husbands and wives, mostly in their fifties, found themselves alone. At weekends and during the holidays, the children would come back to their parents, but during the school year there was an 'empty nest' situation in the home. Suddenly the house was no longer filled with the "babbling" (Mrs Stehlíková's expression) that had been so characteristic of the large rural building.

"I imagine you miss that 'babbling' now," remarked the interviewer a little tactlessly. However, another typical feature of the Colourful's approach to life is that they always look on the bright side: "Not at all!" During the year, their friends come to visit – for

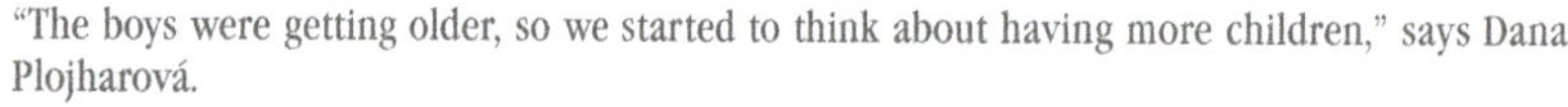

"The boys were getting older, so we started to think about having more children," says Dana Plojharová.

190

example, participants in the seasonal woodcarving courses run by Mr Stehlík or the Volary Reaper scything competition. The Jandas aren't suffering from empty nest syndrome either. When their daughters left home, they "got more livestock". Now they have more time to have a coffee and watch the sheep – a classic example of ecological luxury and the *Lebenskunst* mode.[MK] Particularly in spring, when the sheep are let out into the fold for the first time after winter, the joy of the animals makes for a wonderful sight. "It's great fun," is how Martin Janda evaluates this situation, which contains elements of contemplation.[OK]

After her children left home, Jasmína Vaňková also focused on the joyful moments provided by the animals, the garden and the nearby countryside. She bought a small hut, grows vegetables in the garden and has fulfilled a childhood dream: she has a flock of sheep.

The Plojhars dealt with the pitfalls of empty nest syndrome in a radical way. The family's situation was an unusual one. Mrs Plojharová was beginning to thrive as a graphic designer – her posters were a success. The family business was also prospering. The successful entrepreneurs were slightly concerned that the business could overwhelm them and jeopardize their relationship. When their sons Jakub and Jan left home, they began to think about having more children. It took them two years to make up their minds, and then Rosalie was born. And because "the boys were getting older" and the parents didn't want her to be an only child, Mariana also came into the world. "It was a fantastic decision," said Dana Plojharová.

One thing was unmistakeable during the interview: When the cameraman had the couple sit close together to fit into the shot, they showed no signs of hesitation. It was clear that they enjoyed the physical presence of their partner; they looked and behaved like

↓ "Matěj is the best thing that could have happened to me."

→ During the interviews the married couples (Mr and Mrs Franče) behaved like newlyweds.

newlyweds. What's more, the answers they gave to our questions and the way they phrased them demonstrated that they had talked a lot of things over together. It seemed to me that this feature of the couple's behaviour was new, or certainly more conspicuous than on previous visits.

My academic view of the Colourful's marriages underwent a transformation:[252] While in 1992 I was confident of the strength of their unions and noted the traditional character of family and gender roles,[UK] ten years on I was surprised to learn of divorces in two of the families, the women's enthusiastic involvement in educational activities and their commitment to their professional work. Now in 2015, I can state that the dysfunctional relationships were replaced by other ones after the divorces. The successful 'old' marriages, which had been given more space in their empty nest, acquired a deeper dimension and new intensity. The avowal or reaffirmation of their relationships, verbalized by the women and grudgingly confirmed by the men, appeared spontaneous and natural, as I had become accustomed to with the Colourful. Nevertheless, they still surprised me:

"It's hard to find the kind of relationship you two have," Jakub Plojhar would say to his parents when trying to explain how difficult it was to find a partner.

"Matěj[253] is a luxury," said Marie Matějková at the end of the interview, referencing the subtitle of the book *The Half-Hearted and the Hesitant*: "I have to tread carefully here because you never know what might happen." 'Matěj' shyly added that he would say the same.

[252] I admit to this in the previous chapter on pp. 165–166.

[253] Meaning Petr Matějka; the affectionate use of the modified surname is characteristic.

← An analysis of the Maláč family roles would detect the 'new masculinity' phenomenon (Segal 1990).

↓ The Jandas don't suffer from empty nest syndrome: "When our daughters left home, we got more livestock."

[254] He has both his wife and mother in mind here.

[255] Later waves of the 'influx' had different characteristics (Loquenz and Šimon 2013). The 'amenity migration' was related mainly to the surroundings of cities and was motivated by the desire to live in a healthy natural environment. It would appear that 'amenity immigrants' did not have the same interest in the life of the community as those from the 'influx' of the 1990s. In many cases the children were driven to the town each day by their parents instead of attending the local nursery and primary schools.

[256] A vital question for the future development of these communities, which benefitted from the influx of civic-minded individuals in the past, is whether the children of the Colourful will return to the communities they grew up in after studying and experiencing city life. Their parents would certainly welcome them back. We will examine this possibility in Chapter IV.4.

Václav Frančе spoke sympathetically and appreciatively of women's work:[254] "I always say that my wife must have some kind of engine because she manages to do the washing, tidy up, do the shopping and cook as well as look after a child. The way those jobs are linked together is seemingly incoherent, but if you add it all up – because I wrote out a timetable and the list goes on and on, from 6am to 6pm, it's a twelve-hour shift. Why do they call it *mateřská dovolená* ['maternity leave', literally 'maternal holiday' – translator's note] when it's anything but a holiday?" Mr Frančе willingly participates in the non-traditional division of labour and approaches it rationally: "I removed the iron aspect from the ironing."

Anička Maláčová, atypically recently married for our sample: "I don't even say anything in front of the other mums – they're starting to guess how nice he is, and they'd be jealous because their husbands don't put up with all of that. He regularly has ten- or twelve-hour shifts, but even so he'll get up at night for the kids. I'll have had enough of it, but he'll get up for the kids."

Serving the community: neighbours, a mayor, councillors, civic activists, an entrepreneur

The children leaving home has been the most significant change in the lives of the Colourful over the last thirteen years. Meanwhile, one constant has been their involvement in the life of the community – *vita activa* in the *homo agens* mode.[OK] Social commitment and the opening of the families to the wider community remained characteristic of the Colourful.

In 1992 some of the Colourful were part of the so-called 'influx', a peculiar phenomenon in Bohemian and Moravian villages which was a reaction to the political situation and adverse environment in large towns and cities – a reaction which was linked to concerns about children's health. The influx wave of the early 1990s was relatively strong and had specific roots (Librová 1996, 1997). Among other things, it was connected to the idea of a new start in life which formed as the communist era came to an end.[255] In many cases the 'influx' became an important source for the revitalization of a village's social and cultural life.[256]

The Colourful's contribution to life in their communities can be seen on many levels: as neighbour, councillor and civic initiator of community life, but also entrepreneur.

In 1992 the Plojhar family said that their secluded home at Podhradský farm attracted "more people in one week than our flat would in a month." Naturally, these visits were motivated by curiosity. The Stehlík and Janda families, who are neighbours in the semi-seclusion of Volary, are also visited by friends – no longer out of curiosity, but for the social events they regularly hold. However, there is also a bond in the everyday life of the Stehlíks and Jandas which has elements of the kind of true neighbourliness we find in village life. People exposed to the harsh conditions of nature are dependent on the mutual help offered by neighbours. In the early 21st century, in an age of cars and mobile phones, that help might not be a matter of life and death, but it is important. This certainly applies to the children's home schooling. The mothers from the Stehlík and Janda families take turns to teach the children and can count on a helping hand if one of them is ill or otherwise indisposed. It is taken for granted that families will help each other out with the farming. The Jandas told us that they would be able to have a week's holiday because their neighbours the Stehlíks would look after the sheep. The everyday contact between these families also has a rural character far removed from urban soirees. "We don't actually visit each other regularly, maybe once a week for a coffee. It's more that we'll bump into each other in our wellies with our wheelbarrows... Then we'll stop and can happily chat for an hour."

However, good neighbourly relations are not restricted to helping out on the farm. The example of the Franče family, who live in Kaplice, shows that living in a block of flats doesn't mean you have to be cut off from nature or from other people. "It's fine in the flats here too, we all get on well. Next to us is Mrs Řimnáčová, who's like a gran to us, even though we're not actually related. (...) I remember good neighbourly relations when I was a boy. A block of flats isn't the most supportive environment for them,[257] but we just like her."

We characterized the first stage of our research into the Colourful in the early 1990s with the slogan: "We looked for modest people and we found mayors." This was obviously shorthand, but if we were to add – "We found mayors and councillors" – we wouldn't have been far off the mark. Alongside the low consumption of material goods, another common denominator for our otherwise strongly heterogeneous group is their participation in community life. In the interviews from 2002, some of the councillors had ruled out standing for re-election, or at least had doubts about it, because their experience to date had left them feeling disappointed, if not world-weary. However, we found them on councils again in 2015.

257 It should be noted that the Franče family are not reconciled to living in a block of flats; they own a piece of land called the 'Gorge' which is important to them and where they are building a treehouse. Characteristically, they have not fenced in the land.

The Frančes have also established good relationships in their block of flats. The plate from the cakes will have to be returned to Mrs Řimnáčová.

Mayor Miroslava Válová is unmatched for her vitality and altruism.[PK] She was born in South Bohemia and lived in Prague, but her experience of city life led her, as was typical of the 'influx', to leave in 1983 and move with her family to Olbramov, a tiny village in West Bohemia. At first she lived the country lifestyle: she had a cow, sheep, rabbits, geese, ducks, hens, guinea fowl and, of course, a large garden with a greenhouse. Her household was literally self-sufficient. This changed when she was elected mayor and took on more and more functions. She became chair of the civic association Pomozme si sami (Let's Help Ourselves) and is the main organizer of activities in the Konstantinovy Lázně microregion. These days Mrs Válová no longer owns any farm animals; she shows visitors her overgrown garden with humour and even a certain pride.

Miroslava Válová's approach is exemplified by the way she spends Christmas Day. While families in other villages are busy with preparations in the kitchen, wrapping presents and decorating the tree, in Olbramov it is different. Although not a practising Christian, Mayor Válová organizes a religious community event: "The whole village gets together on Christmas Eve and we go to the chapel,[258] where we sing and wish each other all the best before going home to celebrate Christmas."

Although the mayor's account of life 23 years on was compelling in every respect,[259] its culmination was the story of her pension. During

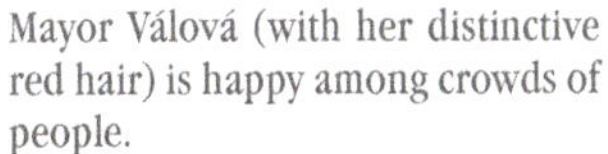

Mayor Válová (with her distinctive red hair) is happy among crowds of people.

her long years of service to society, she was not interested in calculating her pension and the state assessed it at 5,600 crowns [ca. 220 € – translator's note] for a part-time mayor. The otherwise resilient Miroslava Válová admitted: "I did feel sorry about that."

It might almost be said that those among the Colourful who were not councillors usually took it upon themselves to initiate and maintain a buoyant social life in the community. Here are four examples:

Lia Ryšavá continues to promote culture in Miroslav, which she used to do with her late husband Tomáš. They established the tradition of the Miroslav fairs, which are popular in the wider area and contain unusual elements – for example, the line-up includes classical music concerts. Thanks to their dedicated and systematic work, the Ryšavýs managed to transform an ugly communist leaflet founded in 1958 into an attractive, engaging and popular bimonthly *Zpravodaj* (Newsletter), which has an established structure and regular sections. From a sociological perspective, it is significant that Mrs Ryšavá was able to hand over the organization of the fairs and the bimonthly newsletter to the town. Her only involvement in the running of *Zpravodaj* these days is as a member of the editorial board. This is the best possible outcome for civic initiatives: rather than becoming exhausted and embittered through thankless work on behalf of one's fellow citizens, the project which is already up and running is passed on to the appropriate state-designated institution. The time this has freed up has allowed Lia Ryšavá to become involved in the association Stromy v krajině (Trees in the Countryside).

In our earlier research, we came across Mr Roman Kozák as a councillor in Volary and an amateur dramatics enthusiast. He is now a sought-after speaker at weddings, funerals and special anniversaries. He is also involved in organizing cultural life. "There was a pastoral assistant here just now – the sexton. I work with the church to put on concerts there. I recently came up with the idea of opening up the church tower to the public."

Dana and Bob Plojhar set out on an unusual path of civic activism. They clearly have an aptitude for business; in terms of our classification, they tend towards teleologically based behaviour, while most of the Colourful are more *deontologically* orientated.[MK] In 1992 the Plojhars were living at Podhradský farm, where they were planning a waterfront pub and a B&B. At that time they had overestimated their financial and other reserves; however, an even more decisive factor for their business activities was that Bob inherited the family cardboard

[260] For more on the relationship between voluntary and intentional modesty, see *The Half-Hearted and the Hesitant* (Librová 2003, pp. 27–31).

[2261] Guerrilla gardening is a form of social protest which aims to improve public spaces. At the same time, it questions the traditional concept of ownership rights. Guerrilla gardeners plant flowers on land they don't own and try to rejuvenate unsightly, neglected areas (Boček 2009).

business in the city of České Budějovice. Today the company has 70 employees. The Plojhars believe it is important to offer people work and opportunities to earn money, and they employ a gardener and a cleaning lady. This is certainly very different from their intentional poverty[260] at Podhradský farm in 1992.

The Plojhars try to link social and environmental perspectives in their business and their private lives. In their company they promote environmental education and practices, and the company vehicles run on natural gas. Those around them are no doubt aware of how the boss and his family behave outside of the firm. Mrs Plojharová cycles to work as often as possible and their children usually take public transport. People have also probably noticed the Plojhars' attempts to beautify České Budějovice's public spaces, even though these sometimes take the form of 'guerrilla gardening'.[261]

At first glance Vladislav Bukáček's choice of business and lifestyle appear radically different. In the community of Neratov in the Orlické Mountains, he established Kopeček, a sheltered workshop for people with varying degrees and types of disability. He also tries to bring together the environmental and social dimensions in his business. He is, of course, in a completely different situation from the Plojhars. He has three children from his second marriage; every day he has to think about how to provide for his family on his salary from the non-profit organization. It is not easy to remain in employment of this kind and it requires a firm belief in the meaningful nature of the work and the positive influence it can have.[PK]

We have abundance now! Ecological luxury 2015

When the research interview turned to the issue of living standards, Roman Kozák, the loyal Czech Railways dispatcher, Volary thespian

Roman Kozák doesn't share people's general dissatisfaction with their standard of living.

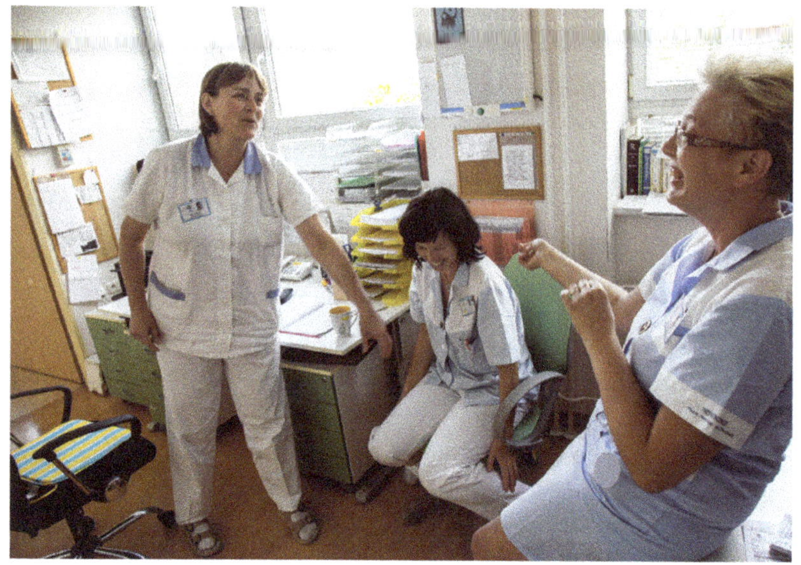

and public speaker, theatrically exclaimed: "We have abundance now!" repeating it emphatically. However, from the arguments he used to justify this and from his intonation I could sense a defiant polemical subtext, as though he wanted to take issue with the trade unionists over their endless demands for wage increases and, at the same time, to draw attention to the fact that it was just for *now*, because who knew what the future would bring.

The Colourful are thrifty, but characteristically they don't keep records of their household consumption. As we have a fairly reliable picture of the consumption of some of the Colourful respondents from another sociological study (Krautová and Librová 2009), and the interviews showed that the structure and volume of consumption had not changed in recent years, I will mention some aspects of their household management without laying claim to any systematic interpretation.

In 1992 two of the households from our group were living on the breadline. This was the case for nurse Marie Bukáčková, a divorced mother of three, when she was working for a non-profit organization. It also applied to the teachers' salaries of Mr and Mrs Ryšavý.[262] However, by 2002 these respondents could claim to be slightly better off in monetary terms. In 2015 Mrs Bukáčková, now Mrs Matějková, had left her NGO and was working as a nurse in a hospital. Her household, like most of the others, was even able to put a little aside. Salaries had improved and it was possible to make prudent purchases – "we

[262] Nevertheless, the Ryšavý family eased the burden on the town budget by using their own resources; for example, they didn't charge for phone calls they made while organizing cultural events.

198

Mayor Válová hitchhiking.

don't buy fripperies". Partial self-suffciency also helped, and although its level was lower than the households had envisaged in 1992, it still played a role in the budget. Even those who were living in blocks of flats grew vegetables, and some of them had greenhouses. In the semi-seclusion of Volary, the Stehlíks and Jandas were raising sheep. Mr Stehlík was able to slaughter lambs, priding himself on not resorting to the buck-passing of those who buy their meat from the butcher. His wife Jitka added in a truly ecological spirit: "The way I see it is if we have a farm, then the farmer is actually replacing the laws of nature... We have land which supports a certain number of animals... so we really buy very little meat."

By far the greatest expense for the Colourful over the 23 years was the never-ending maintenance of property. The majority of the Colourful live in old houses, some in sprawling farm buildings which they have gradually developed an emotional attachment to. For the most part, these are not projects aimed at building something new, but rather a 'caring relationship'[263] – further evidence that the Colourful are more *homines curantes* than *homines fabri*.[OK] The fashion for preserving the original character of a house at all costs runs the risk of historical elements such as a shingled roof or the shape of the windows and the early 20th-century chambranle becoming the backdrop to an otherwise dilapidated building. If an old dwelling is to be given new functionality, it requires significant investment: replacement timbers or repairs to the old ones, sewerage, damp-proofing, water and electricity, heating. The bright side of all this expenditure is the paradoxical fact that there is no money left over for these households to spend on the neophiliac purchase of ephemeral objects – the 'fripperies' referred to above, which will end up in the bin.

In *The Half-Hearted and the Hesitant*, I gave the section which summed up what appliances the households were equipped with[264] the heading 'No escaping the car and computers'. I pointed out that in a small village far from an administrative centre, medical assistance and shops, it was impossible to get by without a car. Obviously, this still applies today, perhaps even more so. "Having a car has made us soft," said Jitka Stehlíková.[265] However, we discovered in our last study that it is possible to resist the imperative to own a car. We found two such examples amongst our Colourful: Miloš Švanda – which was not altogether surprising given his general scepticism towards civilization. On the other hand, it was almost unbelievable that the mayor of a small village community could get by without a car. Not only does Mrs Válová from Olbramov not own a car – she doesn't even have a driving licence. She gets where she needs to go either by hitchhiking or by using public transport; she can also count on help from her neighbours, or from her daughters when they happen to be in Olbramov.

As I mentioned at the start of this chapter, nearly all of the respondents replied by email to the request for an interview I had sent through the post.[266] The interviews then confirmed that computers were of increasing importance in the Colourful households, mainly as a way of keeping in touch via email. In some of the households, the computer was not connected to the internet and was used as a word processor. Most of the respondents were on Facebook, but they all agreed that they disliked its indiscretion and "hadn't been on it" for a long time. In some households the computer had completely replaced the TV screen. Jasmína Vaňková won a television set at a microregion ball but soon gave up on it: "Instead of the telly, I go and chop wood."[OK] For her birthday, the children gave her a wood splitter, which she can't praise highly enough.

The other Colourful also have their own specific, sometimes even bizarre pastimes, most of which can be classed as an ecological luxury due to their aesthetic elements.[MK] Ecological luxury has always been a basic feature of the Colourful, and is even more so now. One new type of ecological luxury is the 'swimming pond' – a fashion that bucks the trend for ecologically harmful swimming pools. Just to give

[266] I must emphasize the word *nearly*. One respondent, Miloš Švanda, was a hard man to reach as he doesn't have a computer, a landline or a mobile phone. "Everyone told me when I started building this house that it would be impossible without a mobile and a car. I bought a phone card for 200 crowns. That would last most people a fortnight, and it did me for six months. I wouldn't like to have to do online banking. I prefer 'counter-money'."

Ecological luxury according to the Svoboda family: logs from their own forest and tomatoes from their own greenhouse.

the reader an idea, the Svoboda family's swimming pond is 1.7m deep and measures 11m × 6m. The visits to their gentlemen's 'pipe club' are a luxury for Martin Janda and Roman Kozák. Dan Maláč gets his luxurious pleasure from looking after the tall apple trees in the old family orchard, while the Svobodas plant oak trees in their wood. Ivo Stehlík has written a book about home schooling, and Vladislav Bukáček writes poetry. On page 183 I show how our respondents' joys in life are perfectly in line with the characterization of the *new luxury* given by H.M. Enzensberger (1996).

The Colourful talk about some of these ecological-luxury situations with the same enthusiasm as a travel agency luring customers with the wonders of exotic landscapes. But it is worth noting the gulf in the concept of what constitutes a *peak experience*: the Franče family delighted in discovering a natural water source in their garden; Ivo Stehlík beat his younger friends in the 800m dash – a race to catch an escaped lamb. One bright, clear day when he was ice-skating on the half-frozen Lipno Reservoir, he was able to see shoals of fish below him responding to his movements. For Jasmína Vaňková, a peak experience is "the view from behind the barn" towards Vlčí hora.[267]

It seemed to be characteristic of the Colourful's basic attitude to the world that even as they relished talking about these luxurious experiences, the conversation seldom touched on holidays. It is necessary to point out that the view of holidays was one of the major changes

According to Jasmína Vaňková, the view of Vlčí hora (Wolf Hill) makes up for all the disadvantages of living in the countryside.

in the lives of the Colourful. In 1992 most of the respondents had exhibited environmentally friendly domicile tendencies,[268] emphasizing that holidays weren't important to them. They claimed that they didn't like to travel far from their homes or that they were happy with a holiday in the Czech Republic. I wrote in *The Half-Hearted and the Hesitant* that in 2002 our respondents were now holidaying in Croatia and had discovered the beauty of the northern seas. Over the next thirteen years, most of the Colourful went on holiday almost annually, though not through travel agencies. A new element was the holiday with grandchildren. But just to keep the record straight, by 2015 we were again hearing statements that testified to an attachment to the local area.[269]

One of the basic characteristics of this lifestyle is the way of dealing with waste. While in recent years most people have been showing their environmental awareness through recycling,[PK] the Colourful don't mention any such virtue. For that matter, they don't put much into recycling bins.[270] Their households are distinguished by *low metabolism*, the limited supply of things. Moreover, one of the Colourful's characteristics is their *faithfulness* to things.[UK] Mr and Mrs Kozák proudly consider themselves to be "scavengers". Martin Janda used to work at a recycling depot, which was an excellent opportunity to bring home objects that had been thrown out and "might come in useful".

The train dispatcher and amateur dramatics enthusiast Roman Kozák offered a passionate defence of old technology: "Those steam trains – that was something that worked for a hundred years... Nowadays it's all just smoke and mirrors! (...) I've got floppy disks for the computer

[268] I find the word 'domicile' a good way of expressing the environmental virtue which is the antithesis of the ecologically damaging nomadism that views leaving the home as freedom.

[269] As the daughter Anička revealed during the research into the children of the Colourful, dragging the painter Svoboda away from his home and studio in Radňovice was no easy task.

[270] The much-vaunted recycling of waste is often a kind of selective disposal resulting from overconsumption. In addition, the energy demands of recycling have to be considered.

Train dispatcher Roman Kozák has no option but to keep up with technology (black-and-white photo from 1992; colour photo from 2015).

[271] Kateřina Štěpánková investigated people's attitudes towards gifts in a representative study of the Czech Republic (2016).

there, everything's screwed up – I don't enjoy having to learn something new all the time." Mrs Iva Kozáková shares the same view: "You learned something and then you spent your whole life perfecting it, but these days you always have the feeling you don't know how to do anything."

The distinctive relationship the Colourful have to things is also reflected in their ambivalent attitude to gifts.[271] This corresponds to their warm relationship with people on the one hand and their dislike of shopping on the other. When the conversation turned to the topic of gifts, strong personalities such as the painter Svoboda openly frowned. Others felt burdened by the situation and explained that they "have everything", "don't need anything" and that it is force of habit which makes people give gifts. Other respondents expressed a kind of contrition: They don't want to go shopping, put it off to the last minute and then feel pressured into buying "some nonsense".

The way our conversation about gifts developed was very typical: at first a clear expression of joy – it's a pleasure to think of a suitable gift to give to someone close to you (with the emphasis always being on the act of giving rather than receiving). However, as the interview progressed, the attitudes expressed became increasingly reticent or even negative. Respondents began to recall problematic situations and difficulties finding a place for gifts; only books were viewed positively. Finally it transpired that "if it has to happen" then the best gifts are home-made ones. At the end of the conversation about gifts, Mrs Maláčová praised the fact that "Dan doesn't like gifts, but he doesn't spoil everyone else's fun."

A note about the Colourful in their role as parents

It is necessary to say something about the Colourful in connection with their parental role. A paradoxical situation kept occurring in the interviews: The subject of the children repeatedly came up in various contexts. For example, it was clear from the interviews with the Stehlíks and Jandas that over the years home schooling had affected the life of the entire household. At the same time, though, when we asked them about their approach to bringing up their children, we were told to ask the children directly: 'He/she is the best person to tell you that.' The results of the interviews with the children will form the content of the next chapter. For now I will just mention two features from the way of bringing up children which proved to be significant from the parents' perspective.

The Colourful are not the kind of parents who assess each domestic situation from an educational perspective. When asked about the approach they took to raising their children, they replied that they had basically given their children free rein since they were young. They trusted them. They more or less intuitively relied on the fact that the domestic situation would be an inspiration for them, that the children would be more interested in family outings than competing with their peers over brand-name clothes, and that there would be no need for drawn-out explanations or ramming the point home. The Plojhars' daughters don't have mobile phones and don't seem to miss them. Their mother is aware that when a child has a mobile phone it gives parents more control and a sense of security, but she thinks it's more important to encourage responsibility and independent decision-making in the children.[MK] Moreover, the schoolgirls will have to earn the money for a phone working in the family business, just as their brothers did before them.

One remarkable phenomenon, which in the view of parents such as Lia Ryšavá has helped to shape the responsible attitudes of her now grown-up children, was the difficulties the families had to contend with:[MK] living with aged grandparents, the presence of illness and death within the home and relative material deprivation. According to Petr Matějka, children who live with economic limitations develop a kind of 'safety valve' in terms of their expectations and demands. It should be borne in mind, however, that this might be an idealized image and interpretation. There were hints in the interviews that the children were not overly enthusiastic about every aspect of the family's lifestyle; the mother's involvement in organizing community life undoubtedly placed a strain on family relationships. Will we find out something more definite in the next chapter, which presents the results of the interviews with the children of the Colourful?

Most of the respondents believe that things have gone well and that their children have been able to withstand peer pressure from their schoolmates when it comes to fashion and technology. On the other hand, some of the parents acknowledged the importance of external factors which weakened the influence of their upbringing and strengthened modern individualist tendencies.[PK] The parents would say with a sigh, "at their age we already had a family". But now that the children of the Colourful are adults, their parents feel that they can no longer intervene in their development. In any case, there is no sense in drawing any conclusions. As Jan Svoboda remarked, "everything is in flux".

Views of the ecological crisis and their own lives

An important aspect of our interviews was the environmental situation. We posed the question in general terms, but the Colourful mainly related it to the place where they live. They are keen observers and pay attention to detail.[272] Most of the respondents think that the ecological situation is relatively good at a local level as there are more trees in the area due to the success of widespread planting. Lia Ryšavá mentioned the encouraging results from Miroslav's association Stromy v krajině. The characteristic tendency of the

Lia Ryšavá is an elder at the Protestant Church in Miroslav.

Colourful to relate environmental questions to the local situation was also evident in the heat study (see the TEXTBOX in this chapter). Jan Svoboda demonstrated a sound knowledge and interpretation of natural history when he observed and discussed with ornithologists how the numbers of smaller birds around his farm were being decimated by the growing magpie population.[273]

The most sceptical of the Colourful, Mr Miloš Švanda, also spoke in his characteristic style, blending casualness with urgency, about the local situation, which points to the poor state of the planet: "I really think something's changed. I see it at work – when it's supposed to rain, it doesn't rain. The weather's changed dramatically over the past fifteen years. In Třebíč the prevailing winds used to be from the north-west, and now they're from the south-east and they don't bring the rains. A total change. My gran used to always say: 'It's coming from Domamil. There'll be a grand storm and it'll chuck it down.' It rained regularly – that was the tail end of the Pošumaví storms that would come down here from Javořice and then get sucked up by the Moravian-Silesian valley."

Our respondents don't have an optimistic view of the global environmental situation. Sceptical opinions were already apparent in the research from 1992 and 2002, but these have become even stronger in 2015. At the same time, the Colourful do not just resort to maladaptive coping strategies. Their environmentally friendly behaviour and their work for the community have the character of an adaptive defence – altruism.[PK] The environmentally conscious businessman Bob Plojhar has no doubts about things changing for the better and sees this as the result of an internal locus of control. In connection with future developments, he even said, "I'm enthusiastic". He explained his enthusiasm in the spirit of *ecopragmatism* – environmentally friendly technologies are being developed and introduced. He sees it clearly in the orientation of the Prague Business Club, which according to Mr Plojhar is very pro-environmental. The important thing is for these changes to be reflected clearly and quickly in the pricing of products.[MK]

Most of the Colourful do not share this technological optimism. We heard the common metaphor "humanity is cutting the branch from underneath itself", but also a more original statement from Jan Svoboda: "it's like trying to create order in a sty," pointing to the futility of individual efforts and the power of the *external locus of control*.[PK] During our interview, Ivo Stehlík posed the question of whether we were "on the brink of collapse or already collapsing".

In two of the interviews which touched on prospects for the future, I was struck by one word which I had not heard during the previous decades in any of the research into attitudes and opinions, and which we don't hear in everyday communication either. That – fortunately – almost forgotten word is 'war'. My colleague Lukáš Kala, who conducted the interviews in South Bohemia in the spring of 2015, recorded the following in his field notes, appending a sociological comment: "The word 'war' came up (an interesting transformation of the narrative)." Miloš Švanda put it less academically but more succinctly: "The war over colonies was replaced by the war over mineral resources – maybe the third one will be over water and food." "How do you reconcile yourself with that?" I asked with the theory of defences in mind. Mr Švanda turned to affectualization, a kind of masculine version of environmental grief:[274] "I take a dim view of Europe. It's like the fall of Rome, its final years. I wouldn't like to be young now – that's my consolation. I would hate to have to start over as a fifteen-year-old."

Without it even being a subject of our interview questions, the respondents distinguished between the state of the environment, worrying environmental trends and their outcome. This distinction was conspicuous among several practising Christians.[275] They have no doubt that things will work out in the end. When I asked them how they envisaged this happening, they replied simply: "I'm a believer." Daniel Maláč, a churchwarden and father of three, distinguishes between the present perilous state and an optimistic future. Characteristically, his comments are connected to observations of his surroundings: "There

[274] In Chapter III.1. environmental grief was presented as more of a female attitude.

[275] Two of our respondents are staunch Christians of the Protestant faith. As Lia Ryšavá poetically put it, their relationship to the church was as unchanging "as the colour of your eyes". Lia has been a member of the elders for a long time at the Evangelical Church of Czech Brethren in Miroslav. Daniel Maláč is the churchwarden (the chair of ministers) at a congregation in Heršpice near Brno. Most of the Colourful have a faith which religious scholars would term 'something-ism', often pantheistic in character. "I'm searching for my own path," says the train dispatcher Mr Kozák. Mrs Iva Kozáková feels she is a Christian but says: "I don't belong to any church. I'm sort of unclassifiable." On page 176 there is a long quote from Magda Svobodová's answer from 2002. It contains regret over the futile search for a Christian spiritual leader, literally a pastor.

are a lot of wonderful small-scale activities. When we go to the Kyjovka river, there are beavers and small ponds – loads of people have contributed to it, decent mayors or some businessman that'll donate a few crowns. But in global terms I think it's going badly wrong. For example, the soil is seriously damaged and in twenty years that's going to be a gigantic problem that humanity won't be able to cope with. (...) There are tough times ahead for our children – not uninteresting times – but somehow things will carry on. It won't come to extinction. There'll be trouble, but then some interesting solution will present itself." Daniel Maláč doesn't doubt the existence of today's environmental crisis and species extinction. When I expressed surprise that the Lord would allow his work to be spoiled, he said, "That's how we see it, that his work is being spoiled. We're limited by the imperfection of our brains, but this is just one sad episode in the whole great story."[276] This is the kind of attitude we termed 'defence by faith'[PK] in our interpretative keys.

However, there are also some respondents without a background in the Christian faith who see the prospects for the planet as poor but not hopeless: "Anything can happen," says Ivo Stehlík. Vladislav Bukáček says: "When I see how fast things are happening, what it was like twenty years ago... the next generations are really going to be in hot water. (...) I keep on working.[MK] I tell myself that society is dependent on situations such as when someone delivers newspapers, when I go for lunch, when the children come home from school, when we study, when someone repairs the chapel over there, when I plant a tree – that's what civilization is based upon. If we stopped doing all of those ordinary things, we wouldn't get anywhere," adding, "When we're working with those disabled people and I see how optimistic some of them are, I can't let myself be a pessimist."

The Colourful begin to take stock

Thanks to the fact that I was visiting the Colourful for the third time as part of my research, and thanks to the fact that we had also been in direct or indirect contact in the period between the stages,[277] I was able to ask them to give a personal evaluation of their life so far, a kind of existential stock-taking. Indeed, it wasn't even necessary to explicitly ask; the conversation turned to this topic spontaneously and repeatedly; for example, when the respondents were thinking about how their attitudes had been shaped. We were slightly surprised that the respondents who had gone through a divorce spoke about that period in their lives without being prompted.

As we already know from *The Colourful and the Green*, the respondents mainly recalled the influence of their grandparents, who were still alive in 1992 but were no longer with them by 2015. Obviously, the believers emphasized the enormous influence religious faith had on their lives. At the same time, some of the interviews confirmed what we already knew: Some people had been put off religion by their negative experiences with representatives of the church. Often this might be just apriorism. What was even sadder was that not one respondent mentioned a teacher having influenced their values and attitudes.

The most frequent outcome of the Colourful's stock-taking was: "We have no regrets," "I'd do it all over again," or (from the businessman) "I'd like to be here longer cos I like this life." However, the answers we found the most convincing were those which were more nuanced.[278] The Colourful wouldn't be the Colourful if they were all united in how they viewed their lives or if this view did not contain some internal contradictions.

Dan Maláč, a man who described himself as a "cold fish", has a special place in our conversations and the interviewers' memories. We were moved by his open admission. He has still not completely come to terms with the fact that after years of hard work, he ultimately failed as a farmer: "I'd like to let it go, but I guess it's stayed with me – the feeling that I've let myself down, that I didn't make a success of it – my own defeat as a farmer... I don't need to see a psychologist, but I haven't been able to let it go completely."

I make a weak attempt to protest: "You can't look at it that way – the economic conditions were against you," but then I fell silent.

[278] If I disregard the inner emotion which we felt at such moments and assess the interview from a purely academic methodological perspective, it reaffirms the cognitive value of in-depth interviews and their superiority over quantitative methods.

Mr Maláč's yard is testament to the path he's taken in life.

My co-interviewers, Lucie and Vojtěch, also said nothing; the usually chatty cameraman David was quiet; Dan's wife Anička was silent, and there wasn't a peep out of the three children. With an apologetic look at his wife, Dan quickly added: "But it's true that I wanted to make the family my priority – I wouldn't turn the clock back."

But the silence lingered around the table until the tactful and unfailingly kind Vojtěch Pelikán quietly objected: "But you haven't entirely given up on farming. You still do some farming, even if it's on a small scale." The film footage shows how our respondent brightened up again. "It's true that I'm happy I've kept a small piece of land – half an acre. I'll plant some trees here and there. It'll be great to go there and mow something, dry a bit of hay – we might need it when we get the sheep. It's been well looked after, tastefully done. That overgrown baulk is full of wildlife. I hung some bird boxes there... a blue tit in one and a great tit in the other – there was even a shrike. I like that kind of ecological farming – almost."

When asked about how she coped personally in the midst of an ecological crisis, Mrs Jitka Stehlíková answered: "I try to work spiritually. (...) Yes, I want to shine a light on how we might live more frugally, but I don't believe it would mean a great change in global terms. We do some farming; we're linked to the land in some way and that takes its toll. You always need time, energy and money to come up with some kind of revolutionary ideas... Money isn't always necessary, but time and energy are." Her 52-year-old husband added: "What can I say...? Time... We don't know how much we have left – thank God. But it's very, very unlikely that I have more time ahead of me than behind me. And we're not getting any stronger. We're at the stage where you have to ask yourself what you still want to achieve."[MK] In a few sentences, Ivo Stehlík had summed up the situation of the Colourful in 2015.

From the interviews with the Stehlíks, you can sense the wisdom and joy of life beneath Plechý mountain in the Šumava region. See the 'heat study' in the Textbox at the end of this chapter.

A brief comparison with international research on voluntary simplicity

In Chapter III.5. I wrote about why it is difficult to compare the results of our research into the Colourful with studies into voluntary simplicity carried out abroad. The differences we come across need not be factual; they may have been caused by different methodologies. Primarily to give the reader food for thought, I will present the main differences between our findings concerning the Colourful and the information on 'simplifiers' provided by the much-cited quantitative research by

Samuel Alexander and Simon Ussher (2012) from the Simplicity Institute in Australia. These authors view the people who adopt environmentally friendly ways of life as conscious adherents of a grassroots
movement and as a potential political and social force. In their view,
'simplifiers' consciously and deliberately attempt to bring about systemic social change. Unlike the Colourful, they gravitate towards an
ethic of ultimate ends, as defined by Weber.[UK] 67% of the research participants declared that their lifestyle (simple living) had a direct link
to political activities.[279] According to these researchers, it is extremely
important for the Simplicity Movement to radicalize, expand and become part of the social, economic and political mainstream.

This aspiration is not part of the Colourful's lifestyle. Even though it
is also characterized by a high level of social contact, this is not viewed
as part of a political movement. The Colourful are distinctive individuals and their environmentally friendly lifestyle is not dependent on
ideological trends. Although they make a significant and active contribution to civic life, their integration into local structures is not linked
to a green ideology. Nor do they set themselves the aim of spreading
their private environmentally friendly lifestyle.[MK][280] This is partly because they see their influence on developments as weak, if not insignificant,[281] and partly because they are essentially tolerant, benevolent
and undogmatic (Frasz 2005). It is certainly likely that the behaviour
of the Colourful influences those around them, but this process has
more of a passive character: that of setting an example to others. Our
findings were closer to the view taken by Mary Grigsby, another leading researcher in the field of voluntary simplicity, who used qualitative
research design as the basis for interpreting the Simplicity Movement
as a grouping of individuals who are trying to escape the political and
economic system and are not attempting to change it on a collective
level (Grigsby 2004).

The Colourful differ markedly from the respondents in the Australian research in terms of their motivation. 'Simplifiers' are loyal to the
ideology and rhetoric of voluntary simplicity. The prime reason they
gave for their lifestyle was 'environmental concern' (more than 80%
of respondents). Their motive has a strongly teleological character[MK]
and an intellectual radicalness which neither Czech environmental activists nor Colourful-like citizens would commit to. Not once did our
respondents cite an environmental motive, despite the fact that our interviews demonstrated that they think about the connections between
their lives and environmental issues.

[279] Alexander and Ussher believe that
their data allows them to estimate the
number of 'simplifiers' in the world,
or at least in economically developed
countries, as well as the dynamics of
future developments. They take their
lead from the study by Clive Hamilton
(2003) which estimates that in Great
Britain 25% of the population aged
30–59 are undergoing a process of
'downshifting'. Alexander and Ussher
state that there are about one billion
people living in developed countries
and 20%, *i.e.* approximately 200 million, are part of the Simplicity Movement (2012).

[280] In our report on the research twenty-three years on, I mentioned one respondent, an entrepreneur, who had
this aspiration.

[281] They view an external locus of control as adequate.[PK]

282 In public opinion polls, health in general is seen as one of the most important values and motivations for certain lifestyles.

The motives of the Colourful have a distinctly deontological character, a sense of tradition, and are often religiously rooted. Here the latent features of environmental virtue ethics dominate, which incorporate aesthetic considerations and a high level of sensitivity towards nature.[MK] This focus could be observed in real-life situations such as the need to escape the city. In 1992 a typical respondent told us: "I wanted to get out of Prague because Prague exhausted me with its trams, all the people, the hustle and bustle and no time for anything. We wanted to go to the countryside, to nature, preferably to a remote place; we placed adverts: 'preferably remote'." The participants in the Australian research mostly live in cities, and Alexander and Ussher very critically described the lifestyle associated with moving to the country as escapism.

The second motive Alexander's and Ussher's respondents gave as the motivation for their lifestyle was health.[282] The Colourful do not pay any special attention to health, at least not when talking to us. Here I quote the same respondent, who spoke with characteristic exaggeration and humour about how her children's health was affected by the radical change in her family's lifestyle and the move to a rural retreat. "They survived; here the children have natural selection. The weak children can't handle it, they're always ill, and the stronger ones endure. Matěj was born in Prague and he's a weak child, he's always ill, whereas the two younger ones aren't."

A sociological heat study

"There will be hot currents of air coming towards us tomorrow from Africa (from the north-west Sahara). Tomorrow afternoon the mercury in the thermometer will rise to between 32°C and 36°C. The heat will reach its height on Friday and Saturday, when afternoon temperatures will settle between a tropical 34°C and 38°C. So the whole of the Republic will see records being broken for this period."
(Czech Hydrometeorological Institute report, 12 August 2015)

Yvonna Gaillyová was speaking from the heart when she described those days as a "constant heatwave horror".

Our poet, Radek Štěpánek, reacted to the heatwave with expressive verse:
"We will come to curse this sun, / shield our skin from it, / hide our eyes behind dark glasses, / dress in rags and burrow into the earth, / concentrate our poisons in fleshy bodies / and cover ourselves in thorns."
(2018, p. 37)

Were the Colourful also afflicted by weather grief? Their written responses perhaps tell us more about their relationship to the world, their lifestyle and their attitudes than the complex sociological interviews do.

Heat study in South Bohemia

Martin Janda, Brixovy Dvory, Volary:
On the whole, we're fine. And despite the fact that I occasionally have sunburnt eyelids and my arms are dragging down to my knees, there are definitely times when I feel sorry for those poor folk in the city who don't have the shade of trees at their workplace or a river to jump into after work.

Jitka Stehlíková, Brixovy Dvory, Volary:
Generally speaking, it's not as bad here as in lower-lying areas. Since we're almost 800m above sea level, there was just one day when the temperature climbed to 33 degrees, which is a record here – normally it stops at 30. There's lots of dew at night, so the land isn't as dry, and we even had aftermath which dries really amazingly in this weather. The hay harvest, which is our biggest seasonal job, went like clockwork... We take lots of water with us, so we drink it and I sprinkle my clothes with it and soak my hat, and it's great. (If it then suddenly cools down, we fire up the stove, and that's nice too.)

By the end of this long period when it hadn't rained for ages, we knew things were going to get tough – the forests were really dry, the blueberries had dried up and fallen off, the little pond in the pasture where the horses usually go to cool off had dried up, and the water stopped flowing from the nearby well into a historical spring. We had to resort to an alternative source which we don't normally use. That happened for the first time about two years ago for a short time, and the winter before last, which was unusually dry with no snow. But as soon as it rains, the spring starts flowing again (so far).

Iva Kozáková, Volary:

I'm wearing strappy summer dresses all the time. I've never worn them so much. It's a dream fulfilled...

I cycle to the River Vltava to swim. There aren't any boats because there's not enough water.

I walk barefoot. I banged my toe against a rock in the water. There's horrible pus coming out of it. My foot hurts.

I'm always standing up at work. My hip hurts.

My dad has deep-vein thrombosis. He's in hospital.

Heat study in South Moravia and in the Bohemian-Moravian Highlands

Lia Ryšavá, Miroslav:

It's hot. A bit more than we're used to. Thoughts turn to mush like ripe apricots under a wilting tree. It's hot... Occasionally the wind picks up, but it brings no relief. It is dense and hot, the blast from a furnace, the wind from the Sahara, the heat of the flames... It is better not to think, just to switch off... but that's impossible. The only lively things are the wasps, occasionally they even sting. When they sting it really hurts – in this heat someone has to slog away on a motorway, a construction site, in a field. The heat is relentless. It would be best not to think, not to go out, to stay at home... but even there the heat is suffocating, and somewhere on a road, in the dust, in the scorching heat, people search uncertainly for a home – wasp stings only hurt; uncertainty and helplessness, where will it end?

Magda Svobodová, Radňovice:

I had my first big crisis at the end of July because nothing was growing even though I was watering everything. Some of my friends just gave up at that point as they had nothing left to water with. We have a pond and a well, so I treated the watering as a work-out and a challenge at the same time. I felt sorry about the grain and potatoes in the field behind the house. At the same time, I wondered if it would be possible to harvest something under these conditions. (If we are to survive something, it'll be dependent on the crops from the field.) And surprisingly, we probably could survive. The yield was pitiful, but we wouldn't go hungry in winter. I have to take my hat off to the way Mother Nature has it all worked out, that all of the roots (including the grass roots) managed to find some moisture even in this horrendous drought and turned green as soon as it rained.

Marie Matějková, Žďár nad Sázavou:
It really is quite a heatwave and I have to admit that it's hardest to cope with the heat at work, where there's no air conditioning. I feel quite sorry for the patients. It's all right at home – a stone house is a stone house. It's about 24 degrees there if we don't open the windows much. So it's no problem. In the evening we go for a dip in the local reservoir, which used to be a back-up source of drinking water. It's true that the water level is dropping, but it's still clean at the moment.

IV.4. The Children of the Colourful

Vojtěch Pelikán, Lucie Galčanová Batista, Lukáš Kala

Initial scepticism

In the previous chapters dedicated to the Colourful, there were numerous references to the children. In 2002 the conversations began to move in this direction repeatedly and with a new urgency. This was no coincidence – in most of the modest households, a new generation was slowly coming of age.[283] There were crucial questions hanging in the air: How do the children feel about the unusual and in many ways demanding way of life without material comforts? And by extension, how will they feel when they leave the parental nest and start their own family? Will they follow in their parents' footsteps or adopt a critical stance towards them?

The Colourful expressed the hope that their children would ultimately withstand the peer pressure from others *au fait* with the latest television series and dressed in fashionable brand-name clothing. They hoped that as time went by – partly through their upbringing and partly through the influence of their siblings – the children would come to appreciate their parents' values and way of life. At the same time, however, most of them were aware that there was no foundation

The 'children of the Colourful' research team.

for these hopes in their own experience. The opposite attitude was typical for them: They didn't identify with their parents' urban and consumer-oriented lifestyle. They had drawn their inspiration elsewhere – from their grandparents or from traditional ways of farming.[284] The Colourful ultimately admitted that it was hard for them to guess what path their children would take in life, adding that we would have to come back in a few years' time.

We began to take up the Colourful's challenge in the spring and summer of 2015. In the meantime, the three of us – two young male researchers and one female researcher from among Hana Librová's former students – had become attached to the subject she had held such lively discussions with the Colourful about. For us, the children of the Colourful[285] represented a natural choice for developing the original research further. We were curious: What direction had the lives of the children from remote West Bohemian and South Bohemian farmhouses gone in? Had the offspring of the Colourful developed a taste for the benefits of 'ecological luxury', or had they had their fill of a modest life? Did it bother them that they didn't get to go on foreign holidays? Did it bother them that they had to go around in charity-shop and second-hand clothes?

There was one other substantial aspect that influenced our thinking. The period when most of the Colourful set up their households was in many respects unique. The post-revolution atmosphere was redolent with optimism and was conducive to new, life-changing decisions. It was also a time that saw an unprecedented interest in environmental issues. Since then, a lot has changed. Over the years society had grown richer and lost its initial idealism; its level of consumption had also increased – by approximately three quarters in real terms. Like the Colourful, their children had surely not 'stayed put in moving sands' either. What effect might the transformation of the *Zeitgeist*, the spirit of the time, have had on them?

In all honesty, when we picked up the phone and then set out on the field trips, we were sceptical.[286] One reason for this has already been mentioned: the Colourful's critical attitude towards their parents' socialist/consumerist lifestyle. However, there were also other warning signs which we couldn't overlook: for example, the experience of community forms of co-existence.[287] The usual outcome of their situation is for the offspring of the generation that founded the community to lose interest in their parents' radical way of life and move to the city for a more comfortable existence. There they allay the feeling of

[284] See Chapter IV.1.

[285] Although most of them have now reached the age of majority, we have stuck to the designation 'children' in the text.

[286] In a seminar discussion about the forthcoming manuscript, Štěpán Koza suggested calling this part of the trilogy 'The Grey and the Corrupted'.

[287] These were mentioned in Chapter III.5.

having missed out or been socially excluded by turning to consumerism (see, for example, Jones 2011). This apostasy is less conspicuous in communities that are held together by the power of a coherent ideology – typically in the case of religious communities. In contrast, the inability to reproduce the lifestyle in the next generation seems to be characteristic of communities that are oriented ecologically.

The dispiriting experience of many voluntarily modest families – including his own – was summed up in the article *Je skromnost dědičná?* (Is modesty hereditary?) by the environmental journalist Tomáš Feřtek (2011). He writes with undisguised bitterness: "The only thing that can be considered a victory is the fact that [the children – authors' note] do not turn their parents' lifestyle into the exact opposite. (...) Some of them rebel at the onset of puberty, so at the age of about thirteen they'll reproach their parents for this 'voluntary poverty' as a sign of incompetence or hypocrisy. (...) Or the children will put up with that slightly alternative and modest lifestyle seemingly without protest until they finish secondary school, but then they'll look for a job as quickly as possible, because for them the opportunity to earn a regular income is a symbol of freedom and making their own way in life." Feřtek is uncompromising in his judgement: "When I see the children from 'detached' Christian and alternative families at the age of twenty-five disproportionately kitted out with laptops and top-quality brand-name outdoor gear, it's hard for me to interpret it as anything other than compensation. Very often, these orthodoxly modest families achieve the exact opposite of what they intended with their numerous offspring through their careful upbringing."

Other reasons for scepticism emerge from reading the social-science literature. One frequently cited hypothesis is that of the American political scientist Ronald Inglehart (1977), fleetingly mentioned in earlier chapters. Inglehart claims that in economically developed societies there is a structural shift from materialist values focusing on economic and physical security to values such as self-expression and aesthetic gratification or an inclination towards environmental issues. Inglehart concludes that these *postmaterialist values* arise as a consequence of adolescent socialization within an environment of relative abundance; they are not 'inherited' but are derived from the education or income level of the family, and by extension that of wider society.[288] If we take Inglehart's theory seriously, there are unavoidable concerns: How can this be the case in the Colourful families, which have voluntarily given up material comforts?

We could also try searching for answers to whether these sceptical prophecies are borne out in the wealth of foreign literature on voluntary modesty.[289] However, the research effort invested would probably end in disappointment. Despite the seriousness of the question of whether voluntary simplicity has any chance of becoming a more widely accepted norm or is only a marginal and ephemeral phenomenon, little attention has been paid to studying it empirically. Although the historical waves of modesty written about by the likes of Samuel Alexander and Amanda McLeod (2014) never actually met with a strong response within politics or society as a whole, authors from the Simplicity Movement still base their efforts on the assumption that environmentally friendly lifestyles will continue to spread, or write in vague terms about the importance of upbringing and education.[290] The American sociologists Carol Walther and Jennifer A. Sandlin (2013) state that they are among just a handful of authors to have addressed the issue of the intergenerational transmission of lifestyle among *ethical consumers*. The reasons for this surprising state of affairs are unclear; perhaps one contributing factor has been the generally low proportion of households with children in studies of foreign 'simplifiers'. Yet the limited findings we can take away from these studies tend to reinforce our cautiousness. In keeping with Inglehart's theories, they state about their respondents that they mainly come from the (upper) middle class and have rarely followed in their parents' footsteps (*e.g.* Aarts 1999; Elgin 1981; Zamwel *et al.* 2014).

Meeting the children of the Colourful

Our curiosity whetted by the lack of convincing answers we had found in the literature, in May 2015 we set out to see the first of the Colourfuls. Their reactions quickly dispelled our fears. In the eyes of the Colourful, no intergenerational rift had taken place – they spoke of their children enthusiastically, often with audible pride in their voices. They described to us where their children were living and what they were doing, and we talked about their partners and in some cases even about grandchildren. However, when we asked for more details, they replied: "You'll have to ask them directly." They then provided us with an email address or phone number for most[291] of the children and promised to put in a good word for us.

Perhaps also due to parental intervention, we were met with the kind of friendly reception we had grown used to with the Colourful. Only three of the children declined to participate in the research,

[289] We attempted to get to grips with it in Chapter III.5.

[290] Surprisingly, this inconsistent approach also applies to the *degrowth movement* (which the Simplicity Movement is often considered part of), which has strengthened in recent years and seeks a comprehensive overhaul of the socio-economic system – and yet the potential for spreading ought to be a key issue for the movement.

[291] In four cases they expressed the wish that their child be left out of the research for family or health reasons.

Who would recognize these nice young men as the little Plojhar boys from the photograph on page 131?

while the rest invited us into their homes. We can only speculate about the reasons of those who refused. However, there are grounds to fear that a reluctance to look back on their childhood or scepticism about their parents' way of life may have played a role in the decision. On the other hand, we obtained a lot of information about these three during the interviews with their parents or siblings and there was no indication that they differed substantially.

Their approachability was not the only way in which the children of the Colourful resembled their parents on initial contact. They also turned out to be characterized by the quality that Hana Librová termed 'self-stylization turned upside down'.[292] On the telephone the respondents told us: "You're welcome to come, but you'll probably be disappointed. There isn't much that's green about us." The playing down of environmentally friendly features of their lives also occurred during the interviews. Our fears that they would try to please the researchers by styling themselves as 'voluntarily modest children' were not borne out at all. The children came across as authentic and answered thoughtfully. There is another reason why we probably need not fear misrepresentation – one which regularly became a source of amusement to both sides. Although most of the children knew of our research, they admitted without embarrassment that they had never read the books their parents appeared in.

In the end, we carried out 21 interviews over a period of six months with children who had reached the age of majority. We usually went to see them in their homes; in six cases their partners joined in. The conversation revolved around similar topics as with the Colourful. We dealt with everyday issues connected with economic security, transport, shopping and leisure activities. We were also interested in where our respondents were living and why, and the reasons why their household looked the way it did. We also touched upon their attitudes to politics, religion and environmental issues. One new feature of the interviews was the special attention paid to recollections of their childhood and adolescence. The interviews were interspersed with excerpts from videos of the parents filmed in 1992 and 2002. The point of these was not only to refresh our respondents' memories, but also to confront them with the accounts given by their parents at a time when they were often a similar age as their children are today.

The average age of our respondents at the time of the research was 29 years old; the youngest was 20 and the oldest 44 years old. A relevant factor in terms of interpretation is that women significantly predominated, the ratio being 17:4. As with the parents, this group is colourful in the true sense of the word. We also came across some strong personalities in the next generation; even during the interviews with the youngest respondents, we were often struck by their considered attitudes, lack of ideology and sense of responsibility.[UK]

At the level of a basic description of the group, its typical features left us with the impression that the children had indeed set out on a different path from their parents. Although almost none of them had spent the majority of their childhood in a large town or city, 14 of them were now working or studying in one. What's more, unlike their parents – whose focus was on a rural, traditional way of life – they were of the opinion that city life suited them. The Janda and Stehlík girls had left the remote location of Brixovy Dvory in the foothills of the Šumava Mountains to study or work in Brno and České Budějovice. Jasmína Vaňková and Miroslava Válová were also left alone in their remote West Bohemian farmhouses most of the time – three of the children had gone off to Prague, Matěj Vaněk was working for an IT company in Ostrava and two of the daughters were actually living abroad; the one who had remained closest to hand was Mrs Válová's youngest daughter Marie, who was living in Plzeň, almost an hour away. On the other hand, the children often came to visit their parents and many of them had not ruled out the possibility of moving back to the countryside one day. The brothers Jan and Jakub Plojhar, both living with their partners in Prague, spoke nostalgically of their early childhood in the remote Podhradský farm, situated on the bank of the River Vltava below the ruined castle of Dívčí Kámen [literally 'Maidstone' – translator's note]. "Every time we go for a trip around Budějovice, we end up at the Maid," Olga Plojhar Bursíková commented with amusement.

While most of the parents were not attracted by a university education and preferred to engage in manual activities, often farming, almost all of the children (16) were studying at, or had already graduated from, a university or college. Even among the children, we did not find a penchant for exact, scientific or technical disciplines; a focus on the humanities and social sciences prevailed. One significant difference from the parents lies in the approach to employment: professional fulfilment was a more important value for the children. Hannah Arendt would probably interpret this as a move away from the *homo curans* mode.[OK] Even among the children, however, there was still an inclination towards less financially lucrative occupations, especially the caring professions such as special needs teacher or physiotherapist. Like their parents, they were not usually fixated on career success. It was not rare to find examples of professional continuity: the son of the painter Svoboda, Ondřej, had found his own form of artistic expression in blacksmithing, Anežka Bořilová (formerly Bukáčková) was working in health care like her mother, and Jakub and Jan Plojhar were helping to run the family stationery business. In some cases, the children had built on their parents' work in a distinctive way. Both of Lia and Tomáš Ryšavý's daughters were energetic organizers of community life in the place where they lived – Kateřina Fučíková played an active role on the Protestant council of elders, wrote a local chronicle and, along with her husband Jaroslav, was keeping the local

[293] An expression of the *homo curans* life mode.[OK]

Sokol gymnastics organization and culture life going, while Daniela Bednaříková had set up a maternity centre at the Protestant church and, together with her mother, was organizing the Night of Churches. Eliška Frančová was carrying on the handicraft tradition cultivated in the Euro-Indian family. She went to art school, has made leather goods such as a saddle for disabled children and initiated a family project to build a treehouse. The daughters of the mayor Miroslava Válová also rushed to lend a hand when their mother was organizing one of her many social events.

It is usually hard to distinguish the influence of the 'spirit of the time' from other factors. However, there is one respect in which the approach that is typical of the children's generation manifested itself convincingly: demographic behaviour. Although most of the children had reached the age when their parents – who tended towards the model of a traditional family with a large number of offspring[293] – were already raising them or their siblings, they themselves had put off starting a family. Only four of them were married; another four were engaged and eight were in a long-term relationship. Five of them had children of their own. A clear divide is apparent here. While those born in the 1970s were generally quick to start a family, respondents half a generation younger were influenced by having grown up in the conditions of a population trend known in sociology as the 'second demographic transition', which is characterized by a decline in birth rates in rich countries (Rabušic 2001). These structural conditions further reinforced the contrast between the relatively well-established households of the older respondents and the often more student-like households of the younger ones.

They don't buy things they don't need
The basic characterization of the children of the Colourful suggests that the generational change undergone by Czech society from the early 1990s onward has left its mark on them in many ways. Indeed, it seems that the *Zeitgeist* has had a greater impact on them than on their parents; they are influenced more by the life strategies of their peers. Is this also the case with the issue of consumption – an area in which the Colourful were quite distinct? Do the children also prioritize the non-material dimension of life over higher earnings?

When characterizing the children in terms of consumption, we have to make this sceptical admission: If we were to adhere to the criteria used to select the Colourful in 1992, most of the children would

↑ Ondřej Svoboda has inherited his father's artistic talent, but he has found a different outlet for it than painting.

→ Eliška Frančová was the driving force behind the family treehouse project. Here she poses by a future outhouse woven from hazel rods and willow wickerwork.

probably not have been included in the sample. One possible exception is Lucie Mantelová, the eldest daughter of Olbramov mayor Miroslava Válová, with her modestly equipped household, her small but – as she stresses – meaningful income from her freelance work in the field of regional development and her undisguised distaste for shopping: "I really pride myself on making do with little."[294] In general, however, the children have not taken the kind of dramatic steps their parents did at approximately the same age. In terms of the volume and structure of consumption, their households are far less distinct from mainstream ones. They are not so resistant to the consumerist spirit of the time; they purchase more goods and are more involved in the formal economy. In a way, the retreat from environmentally friendly elements is reminiscent of the process of 'becoming half-hearted' which most of the parents have undergone over the course of 23 years. However, even in comparison with the parents' households today, environmentally friendly features have been further weakened among most of the children. Take, for example, the Jandas. A major topic of conversation during each visit to this household has been the dishwasher. While the parents are still resisting the temptation to make the housework easier by buying one, their daughter Tereza has a much more relaxed attitude

[294] To what extent might her approach have been affected by the fact that she and her family have been living in the ecologically progressive country of Germany for years?

²⁹⁵ Of course, the fact that the deviation from the parents' lifestyle was only stated in weak terms can also be attributed to the circumstances of the research interview, which may have strengthened family loyalty.

²⁹⁶ Viewed sceptically, environmentally friendly shopping preferences can also be considered an expression of maladaptive defences, especially diversionary strategies.^{PK}

Lucie Mantelová, the eldest daughter of Olbramov mayor Miroslava Válová, lives in Germany with her family. However, she regularly comes back whenever her mother is organizing one of her many events – in this case the Chlebovomáslové slavnosti.

to buying appliances: she has a number of them and appreciates the fact that they leave her with more free time. "The dishwasher – I love it," she says. "For me, time is a precious commodity."

On the other hand, Tomáš Feřtek's gloomy prophecy about our respondents being "disproportionately kitted out" doesn't seem to have been borne out in any way. In contrast to the Colourful's own experience, none of the children have openly revolted against their parents,[295] even though – viewed dispassionately – this continuity is often more about verbal appreciation than everyday practice. Nor did we get the impression that the children were compensating for something. Although their households might not typically be eco-friendly in character, most of them are distinguished by a clear non-consumerist orientation. The children characteristically avoid loans, although they by no means belong to the propertied classes and only a few of them have a household income that exceeds the Czech average. The households have still been furnished with restraint; "we don't buy things we don't need" was a much-used phrase. The children are happy with their standard of living. "I actually think I have more than I need," says Anežka Bořilová. We heard similar statements frequently.

The children are definitely not overly attached to fashionable styles or brands. They resist the neophilia reinforced by advertising; Marie Válová sums it up well when she speaks of an "innate longing for the old". There is even evidence of faithfulness to things^{UK} when it comes to technologies such as laptops or mobile phones that quickly become obsolete. In connection with this, one episode in the household of Zuzana Vávrová, who lives with her partner and two children in a small house near Třebíč, was particularly noteworthy. Her father, Miloš Švanda, always reliably makes an impression on our students with the way he consistently opposed buying anything new in 1992. This aversion was exemplified by his jacket, which he had had since he was fifteen. When we showed Zuzana an extract from an old interview, she exclaimed enthusiastically: "I've still got that denim jacket of his – I still wear it!"

In general it can be stated that, as in the parents' case, the children's attitudes also bear the hallmarks of what Erazim Kohák calls 'selective demands'. When it comes to shopping and eating habits, our respondents are inclined towards greener alternatives.[296] Although there are only a couple of vegetarians among them, most of them are "not really into meat", avoid big shopping centres and routinely shop in markets or second-hand shops. "It's a lot more fun and it's original," is

Miloš Švanda is a renowned gardener. His daughter Zuzana Vávrová says: "I can just about handle growing tomatoes." What kind of gardeners will Zuzana's two daughters turn out to be?

how Anna Svobodová describes her reasons for buying second-hand clothes. Naturally, these environmentally friendly features are more noticeable in some cases than in others. However, at the very least, each of the children seemed to be endowed with – in the words of Petr Matějka – a "safety valve".

Dave Horton, a sociologist from Lancaster University in England, has looked at the role that material objects play in the lives of environmentally oriented individuals (2003). In his view, two of the key indicators of a non-consumerist orientation are the television and the car. Among the children, the former is usually absent or is watched only occasionally: "Putting the TV on during the day, apart from at the weekend, seems almost like blasphemy to me," says the Ryšavýs' eldest daughter, Kateřina, a teacher at the village school in Prosetín in the Bohemian-Moravian Highlands. The same goes for her sister Daniela, who has taken up residence in the attic apartment of the family home in Miroslav with her own family: "For a long time I didn't even know there were programmes on in the morning." The relationship to the car is similarly reticent. In 1992 only around half of the parents owned one, despite most of them living in a remote location. With the children the situation is similar. Nearly half either have no car – and in many cases not even a driving licence – or did not have one for many years.[297] Perhaps more important than the simple fact of ownership is the way they are used. Most of those who have a car still prefer to use a bike or public transport for everyday journeys to work or to the shops.[298]

[297] It should be noted that these are often students who – in spite of their current declarations that they are not even planning to buy one – might eventually end up getting a car after all.

[298] In the case of both the television and the car, a significant role may be played by the specific attitude of today's twenty- and thirty-somethings. Studies in Western Europe and the USA have shown that members of this generation increasingly regard the car as less of a status symbol, frequently do not even have a driving licence and favour city-centre living and information technology (Dutzik *et al.* 2014). There has been no similar research in the Czech Republic, but more narrowly focused studies suggest that this phenomenon has also made inroads here (Šefara *et al.* 2015; TNS 2016).

Although eco-friendliness is not typical for the household of florist Františka Bukáčková, it became apparent several times during the interview that she too is endowed with a consumerist 'safety valve'.

No mayors to be found among them

Even more than for their patterns of consumption, the Colourful stood out for their engagement in local politics and civic associations.[OK] Among the children, similar tendencies towards behaviour that we would assign to the *homo agens* mode appear less frequently. The inclination to become involved in civic life persists most strongly in families for whom this dimension was characteristic. The reticence of the children may well have been influenced by having witnessed their parents' civic disillusionment. Perhaps an even greater contribution to this state of affairs has been made by external factors, especially Czech society's growing scepticism towards public service over the years (Hadler and Wohlkönig 2012).

In general it can be said that the children's activities, unlike those of the parents, do not take the form of direct participation in political affairs. "They talked me into it; somebody's got to do it," is how the only councillor, Marie Válová, daughter of the Olbramov mayor, sums up her own involvement in politics. The children play their part in elections – virtually all of them go to the polls. Their voting patterns vary, but like their parents they are most often to the right of the political centre; environmental concerns are expressed by repeat voting for the Green Party.

Typically, the children don't engage much at a society-wide level.[299] They don't campaign; they are not burning with the desire to change something. We could take a critical view of this, seeing it as an expression of passivity resulting from the spirit of the time. Or could their attitude be based more on a pragmatic belief in an *external locus of control*?[PK] Although the children are sensitive to environmental problems,[300] their interest in them tends to be a secondary one and they usually distance themselves from the environmental movement.

As a counterbalance to their relative half-heartedness when it comes to society-wide and by extension global issues, the children's way of life displays creative and active features at a local level which could be considered an expression of *sublimation*.[PK] We have already mentioned how the daughters from the Ryšavý and Vála families have carried on with the civic work of their parents. Some of the other children have also found a role for themselves in non-profit organizations and in the local community. They help to repair monuments, organize after-school clubs and plan Scout camps. With a few exceptions, more radical events such as blockades or Critical Mass bike tours are not their style. On the other hand, they support activities of this kind or

[299] Though exceptions can be found here too: Jakub Plojhar has become involved in the initiative Reconstruction of the State, which lobbies politicians to pass laws promoting transparency and better functioning of the state.

[300] The topics of landscape management and soil sealing came up repeatedly. This is perhaps best illustrated by the references to drought which entered the conversations long before this topic began to make it onto the front pages of newspapers during one particularly hot summer (see the Textbox in the previous chapter).

[301] With regard to the question of intergenerational continuity, it is significant that even the grown-up children often spend free time with their parents. For example, the Jandas regularly go rafting or on weekend trips together.

have a more conciliatory attitude towards them than the Czech public (see Librová 2014b).

Paradoxically, in spite of their orientation towards locally anchored activities, the children tend not to be tied to one place.[UK] They move house frequently and routinely operate between several addresses. As a result – and no doubt also because of the predominance of urban living – they are less focused on self-sufficiency. Although gardening is popular among them, their gardens are usually low-maintenance; for them it is primarily a leisure activity. Other features of the parental way of life have also been transformed into less labour-intensive forms. For example, it is rare to find them looking after animals – perhaps only horses used for weekend riding. On the other hand, the children generally devote their free time to activities that are valued in environmentalist circles – apart from gardening, these include handicrafts (for example, making jewellery), artwork and walking or cycling tours. In keeping with the spirit of the time, some of them are tempted by foreign holidays, but they typically plan them independently; they do not like going "at the speed of the crowd", as Eliška Frančová aptly put it. They travel by hitchhiking or in the family car and tend to head for the mountains rather than the sea.[301]

Like the parents and the participants in some foreign studies (Grigsby 2004; Maniates 2002; Schor 1999), the children don't use political or environmental motives to justify their actions.[MK] Their behaviour is not the expression of deliberate self-restraint. Instead, as

Most of the children of the Colourful live in towns or cities and also travel a lot. So Rozálie Jandová's parents have to help her take care of the gelding Monty. She travels almost all the way across the country to see him once every couple of weeks.

with the parents, it is closer to a virtue ethics approach in the *Lebenskunst* mode.[MK] Our respondents are not setting out to change the world, nor is it typical for them to have their life planned out in advance: "If it doesn't happen, it doesn't happen," is how Zuzana Vávrová sums up this carefree approach. The children largely base their behaviour on intuition and (in the broader sense) aesthetic judgement. Their approach to water and energy management is typical: "It breaks my heart when drinking water is used to flush a toilet," says Barbora

Mrkvičková, the eldest daughter of Ivo and Jitka Stehlík.[302] Rainwater management is also a matter of deep concern to Kateřina Fučíková, who moved to the Bohemian-Moravian Highlands from the arid conditions of South Moravia: "When I came here and saw the water flowing off the roofs... and all of it onto the ground, I found it really strange. Even when there's enough water, it seems like a sin just to let it drain away." Similarly, the children don't approach shopping with a carefully considered, clearly formulated attitude. Instead, they tend to find it "annoying", "don't enjoy" it or downright "hate" it.

Green partners

In a number of cases, we were lucky enough to have our conversation with the children of the Colourful enlivened by the presence of their partners. They formed a heterogeneous group in terms of age, background[303] and occupation – they include a plumber, a carpenter, a manager, an IT specialist and a sociologist. There was one thing in particular that struck us about them: The offspring of the Colourful often seemed to have chosen opposite numbers with a similar environmental sensibility. Sometimes this was more a case of latent features which manifested themselves now and then in the course of the interviews: for example, instinctive frugality or sensitivity towards the surrounding countryside. In several cases, however, we came across individuals who were even stricter in terms of their environmentally friendly lifestyle than the children of the Colourful themselves.

A case in point is Jakub Plojhar's wife, Olga. Daughter of the former

Olga Plojhar Bursíková is one of the partners with carefully thought-out environmental attitudes. She compensates for the ecological footprint created by travelling through responsible choices when it comes to shopping and household care.

[302] It is important to note that the Stehlíks have a dry toilet at home. A sceptic like Feřtek would surely have expected their offspring to have a stronger need to distinguish themselves.

[303] It says a lot about the transformation of the *Zeitgeist* that four of them are foreigners.

228

chairman of the Green Party and Minister of the Environment Martin Bursík, she graduated in communication studies and earns her living mainly by acting and modelling. She makes no secret of her interest in animal rights activism. Together, she and Jakub consider the ecological features of their purchases and everyday behaviour in detail; they are capable of walking all the way round Prague with a bag of sorted waste in search of the right recycling bin. She is principled: "I know I'm not going to save the planet by turning off the shower sooner, but I'll do it anyway."

Jakub's brother Jan has also found an environmentally conscious wife. Marie is a sociologist who has researched the issue of *dumpster diving* for expired foodstuffs thrown out by supermarkets. She is also involved in the fight against food waste professionally – she is active in the organization Glopolis and the association Zachraň jídlo (Save Food), where she tries to prevent vegetables being disposed of just because they are 'wonky'. She neatly characterizes her approach as "the burden of knowledge"; it has been shaped by an insight into what actually lies behind everyday consumption: "When I have to go and buy some clothes, I absolutely hate it. I look at all those low-quality, extremely cheap garments and I can't help wondering how much of that the children who sewed them in the factories got. And I know I have to wear something, that I can't go around looking like a tramp... But then I think: 'I don't want it!'" We get a hint of a charming dialogue between partners from a story about the time when Jan was considering buying a tablet: "Marie always makes me think about whether I really need it or whether I might not have enough already. And I have to admit she's right."

The list of partners who are 'green' to some extent doesn't end there. For example, there's the family of Daniela Bednaříková, whose husband takes the bus from Miroslav in South Moravia to his job in a Brno IT company every day. They have a car, but they rarely use it: "When we were setting off for Liberec, Aleš discovered that a bulb had gone in the car. We were already in a bit of a hurry, but he and Vilém went to the garage on foot – because there was no way he was going by car." We weren't in a position to judge what the reason for the mutual affinity between the children of the Colourful and their ecologically oriented partners might be. Could this just be a random impression obtained on the basis of a few interviews? Perhaps we will know more once the children who are currently still single find partners too.

Does a modest life begin in childhood?

Readers of the previous pages may well be wondering: How is it that the initial sceptical prophecy was not borne out? What does the experience of the Colourful and their children tell us more generally about where an environmentally sensitive approach comes from? In order to formulate clearer answers, we would have to go deeper in our research; it would be necessary to take greater account of the psychological aspects and spend more than just a couple of hours with our respondents. However, we can perhaps make an attempt at some tentative answers even now.

One of the keys to understanding our respondents' current lifestyle is revealed when they begin to reflect on their childhood.[304] Almost unanimously, they describe it as happy. Even those who lived in remote locations do not complain of having been isolated. Most of them had a number of siblings and got involved in the parents' cultural and social activities; a number of the families lived in multi-generational households. Although the subject of poverty and otherness naturally came up during the interviews, it was not interpreted as a stigma. None of the children had the impression that they had been deprived of anything fundamentally important when they were growing up: "I don't feel as if we lacked creature comforts," says Johana Vaňková, looking back. She cheerfully recalls the situations her mother vividly described in *The Half-Hearted and the Hesitant*[305] – how she and her brothers used to walk the three kilometres to Olbramov on a daily basis to catch the bus to school: "It was fun. Those afternoon journeys had a way of stretching out for us. When the weather was nice, we'd walk all the way from Černošín, which is about five kilometres, through the woods. We'd stop off for cherries or apples or whatever there happened to be at the time."

Of course, not all of the memories are nostalgic ones.[306] For children from the poorest families in particular, the material situation was sometimes a painful topic. Perhaps the most serious repercussion of the modest family incomes was not the lack of a television or foreign holidays, but fashion.[307] Often, the girls don't like to be reminded of their wardrobe of charity-shop and second-hand clothes. For Daniela Bednaříková this is still a live issue today: "I suffered because of it. We really didn't have money and sometimes we went about like tramps. (...) My mother-in-law would sew something and say: 'That's fantastic. No-one's got one like that.' And I'd say: 'I want what everyone else has got.'" The conversation also turned to fashion frequently when it came to the question of what the children would do differently from their parents. However, in the same breath they mentioned that they were well aware of the reasons for the situation. They knew that it was not so much a situation forced by circumstances as the result of their parents' efforts to live in accordance with their values. They were aware of their parents' satisfaction with their lives and most of them sympathized with their approach: "I saw that this way of life was enriching, that it wasn't shallow," says Kateřina Fučíková, recalling some of the difficult times experienced by a family who had to get by at a subsistence level.

[304] Characteristically, the children – like the parents in their time – did not mention the influence of school. The fact that two of the families opted for home schooling says a lot about their relationship to the school system. In view of the development of institutionalized environmental education, it is surprising that the children did not even draw inspiration for their lifestyle from after-school clubs, the exception in a few cases being the church or an organization like the Scouts.

[305] These can be found on pp. 178-179.

[306] Johana's brothers Ondřej and Matěj did not wish to be interviewed by us. Did growing up in a remote location play a role in this?

[307] This subject came up in *The Half-Hearted and the Hesitant* in the sub-chapter "Do children spell the end of a modest life? A dream holiday", on pp. 171-174.

When landscape architect Anna Svobodová comes across a neglected planter of flowers, she can't resist picking the weeds and withered leaves out of it. She doesn't rule out the possibility of making her home in a village in the future, while her sister Lenka is happy living in the city.

However, the material situation was not the main topic which saw the conversation about childhood and adolescence take on a critical tone. If there was something that bothered the children, it was much more about social or relationship issues. The Vála girls recalled occasionally resenting the invasions of privacy associated with their mother's intensive social engagement. As Marie describes it: "You're fifteen, you're staying in a dorm throughout the week and you're really homesick. You get home to find complete strangers staying in your room because there's some event on. And so you end up sleeping in a sleeping bag in the hall." Age can also play a part in the assessment of childhood. While the Svobodas' younger daughter, the garden and landscape architect Anna, concedes that she would be happy to move into the family farmhouse in Radňovice one day, her older sister Lenka, a doctor of aesthetics at Masaryk University, is satisfied with life in Brno. That may be because she was eleven when she was transplanted from the city and the social ties she had formed. She has mixed memories of childhood: "You've got no-one to hang out with, you're a bit shy... The people here are a bit potato-and-stone."

See you in another ten years?

Critical remarks aside, it is generally true that the children of the Colourful wouldn't change places with their peers. They don't seem to have had much trouble coming to terms with being different when they were growing up. On the contrary, they often present it as an advantage: "I would always read strange books and wear strange clothes. I've been a hipster since I was born. I always had old things because we got hand-me-downs," Marie Válová says with a laugh. "I felt as if that lack of stuff helped me to find my own way. In a way it makes you original," says Anežka Bořilová in a similar vein. In statements of this kind, we can hear an echo of the spirit of the time – a lot of what distinguished the childhood of our respondents is now beginning to be appreciated in certain social circles and is being given the fashionable label of 'local', 'vintage' or 'do-it-yourself'.[308]

One thing that may well have contributed to the children following the parents' way of life is the atmosphere in the family. We don't wish to paint too rosy a picture – the years have brought hard times for many of the families, which in some cases have even ended in divorce. In general, however, the families of the Colourful are characterized by the free and easy approach they take – not only to visitors but also to their own children: "Mum never had some master plan for us," is how this experience – regularly repeated in the other interviews – is summed up by Petra Seidlová, another of Miroslava Válová's daughters, now an HR manager for an electronics retail chain. "At the age of

[308] Western researchers speak of the process of an emerging 'ecological habitus' in the more educated classes (Carfagna *et al.* 2014).

six we'd go off to the woods and we'd come back in the evening – and we survived," says Tereza Jandová. From an early age, freedom was linked with personal responsibility: looking after younger siblings or helping around the house and farm. "Chores first then fun", was the rule with the Frančes. The easy-going atmosphere in the families of the Colourful and the benevolent approach to passing on their way of life is exemplified by one fact we have already mentioned: the children were not forced to read Hana Librová's books as 'required reading'.

It's possible that to a large extent the secret to why the children have not rejected their parents' lifestyle – in contrast to other cases – might lie in this liberal, unorthodox approach the Colourful have. Not only do the Colourful not identify with the environmental or even the 'simplicity' movement – they also have little in common with the visions of ecological intentional communities. Their way of bringing up children is not characterized by strict rules, a unifying ideology or political radicalism. The Colourful's approach is intuitive, benevolent and tends to benefit nature indirectly. This minimizes the risk of their children developing a frustrating sense of being trapped by all kinds of norms. In the end, it would seem that the experience of the Colourful and their children does not actually contradict the ideas of Tomáš Feřtek: "There are probably very different educational models leading to the willingness to live more modestly. Except for one: truly *consistently* [authors' emphasis] applied voluntary modesty."

However, we should not jump to conclusions. It is highly likely that the children's lifestyle will undergo a transformation – perhaps a major one – especially in the case of those who have not yet started a family. As the Colourful ultimately found out for themselves, this event tends to have a significant effect on consumer behaviour; the American sociologist Robert Wuthnow aptly calls children "agents of materialism" (1996). On the other hand, the babies' arrival into the world can also be an impetus for the opposite trend: they can motivate the family to adopt a more modest way of life and reinforce the decision to move to the countryside. It will undoubtedly be fascinating to see what course the children of the Colourful's lives will have taken if we come back to see them in another decade.

V.
Conclusion: The palliative conservation of the swift

No matter how much sadness there may be in this
unfathomable world, it remains wonderful

Ivan Bunin

This book began with thoughts about the swift: in the first TEXTBOX their return to their old nests was the symbol of one keyword – faithfulness. Now, at the end, we shall return to the swift. The issue of their conservation contains the more general dilemma of nature conservation today, which we have discussed in this book and which, again, offers more questions than answers: Should we protect something which is dying out and which, in any case, cannot be protected? Should we continue with the traditional nature conservation which protects wetlands, critically endangered species of amphibians, the remains of primeval forests and endemic vegetation from an ever-encroaching civilization? Should we be vainly trying to prevent the invasion of foreign species? And in connection with our research into lifestyles: At a time of inevitable biodiversity loss, does it make sense to limit our demands for certain standards of living and comfort?

It's impossible to answer these questions with a summary of the research results. Instead, I offer readers a partial reply in an email sent to me by the ornithologist Lukáš Viktora, which resonates with the perspective of the author of this book.

Dear colleague,
This year I am writing again to ask if you have any current information about the swifts. I have a special reason on this occasion: I have mentioned the fate of the swift in the book I am writing, which is why I'm asking once more.
Last year you were concerned that in addition to the loss of their nest sites here, there was also something untoward happening in their African wintering grounds. Has this been confirmed in some way? I have also been thinking that the number of insects has decreased in recent years (decades?). I can't imagine that there are enough insects flying over European urban settlements and perhaps over the countryside in general...
I would be grateful if you would be willing to write me a brief note.
Many thanks and best regards,

Hana Librová
1 June 2016

Dear Professor,
A few years ago the English discovered a 25% decline in the 'aerial plankton' which is the predominant food source of swifts in European nesting sites. If we were to rank the causes of the decline in swifts, the food deficit comes immediately in second place after the loss of nest sites.
As with other long-distance migrants, significant over-flights occur with swifts in winter. They normally remain in the Sahel zone if there is sufficient rain and food, then they

move on to the Congo and neighbouring countries, where they also encounter primeval forests. If their wintering grounds are destroyed by climatic changes or extensive alterations to the land, the long-distance migrants will have nowhere to winter and all efforts focused on protecting their nest sites will be in vain...

Nevertheless, this does not mean that we should slacken in our efforts to protect 'our' swifts. For example, in the first call of the new Operational Programme Environment there is the obligation to protect swifts and bats when insulating buildings, which we managed to have implemented at the Ministry of the Environment of the Czech Republic and the State Environmental Fund of the Czech Republic after four years of effort and pressure...

Wishing you a pleasant evening,

Lukáš Viktora
6 June 2016

During the period when Czech ornithologists have been trying to have laws passed to protect nesting sites, when 'swift schools' have had bird boxes attached to school facades, there has been a decline in the number of insects due to climate change. Is it not then unreasonable, or too late, to convince ministerial officials and children of the need to protect swifts when we know that the majority of the exhausted insectivores will die of hunger? It is clear to Lukáš Viktora: *Nevertheless, this does not mean that we should slacken in our efforts to protect 'our' swifts.*

Such is the inconsistent, intuitive, but everyday position of nature conservationists. In medical ethics it wold be called dysthanasia – a delayed death. The ecopragmatists whom I wrote about in Chapter III.2. emphasize that the damage to the ecosystems is irreversible, that committed nature conservationists should forget about their old-fashioned sentimental endeavours to protect the wilderness because they won't be able to protect it anyway against the impact of devastating changes. Ecopragmatists follow a purposive rationality, as sociologists understand it from Max Weber. In their opinion, it would be sensible for us to forget about swifts and turn our attention and resources to the construction of artificial urban environments where birds won't be missed; they will be replaced by other pleasures.

What were the attitudes of our respondents from our last study – the Colourful and the raptor conservationists – towards fundamental environmental issues? They acknowledge the weakness of an internal locus of control[PK] (Chapter II.3.), but they do not doubt that the individual behaviour is important; some things can be achieved at a local level, such as the planting of trees. On a global scale they all responded with marked scepticism, and only a few of them tried to downplay the problems: "In global terms I think it's going badly wrong. For example, the soil is seriously damaged and in twenty years that's going to be a gigantic problem that humanity won't be able to cope with. (...) There are tough times ahead for our children – not uninteresting times – but somehow things will cary on.

There'll be trouble, but it won't come to extinction. (...) This is just one sad episode in a great story," said the Protestant and former farmer Daniel Maláč, fortified by his faith.[PK]

The Colourful differ fundamentally from the ecopragmatists in that their rationality is value based – to use the terms of Max Weber once more. Although it is always possible to relativize and argue over what the necessary criteria are for nature to prosper, deontological conservatism[MK] (Chapter II.4.) leads the field conservationists and the Colourful to dysthanasia, to try to maintain for as long as possible the world of nature in the state in which we learned to see it as beautiful and valuable. While in medicine dysthanasia is the attempt to keep a suffering person alive at all costs, and which is rightly questioned from an ethical standpoint, conservation dysthanasia is one of the greatest things human culture has created.

What will decide whether the teleological spirit of ecopragmatism dominates in society's relationship with nature, or the traditional conservation approach and its attempt to slow down extinction?[MK] This book searched for the source of the palliative approach[309] in normative theories of sociology, philosophy, history, psychology and theology (in chapters II.1.–II.5.). The roots of palliative nature conservation were found to be in the civilizing process, gradually replacing the primitive features of society, in Christian compassion for the weak, in the remnants of medieval chivalry, in the ethics of responsibility and in virtue ethics with its aesthetic as well as civic dimensions of *Lebenskunst* and *Lebensführung*. In environmental theology, the dysthanasia is present in the process of the replacement of the principle of stewardship by an emphasis on living creatures as siblings.[TK]

The experienced environmentalist knows that for the environmental situation of the 21st century, the comforting notion that 'everything will turn out fine', applicable in human relationships, is extremely unlikely, unsubstantiated and trite. Just as the palliative approach in medicine is based on accepting defeat in the individual's struggle with death, in nature conservation it is based on the defeat of creatures by the aggression of affluent technological civilization. The palliative approach is based on faithfulness to creatures which are in need.

[309] Palliative medicine is care for the sick in the terminal phase of an untreatable illness. I borrowed the related term "palliative nature conservation", from Petr Blahut's book *Mýlit se je božské* (To Err is Divine, 2017). However, Blahut looks at palliative nature conservation in a different way from me. It concerns him that it does not eliminate the causes but only the syndromes of environmental problems.

VI.
Bibliography

AARTS, Wilma, 1999. *De status van soberheid. Een onderzoek naar status en milieu.* Amsterdam. PhD thesis. University of Amsterdam.

ADORNO, Theodor W. a Max HORKHEIMER, 1944. *Dialektik der Aufklärung: Philosophische Fragmente.* Amsterdam: Querido Verlag.

ALEXANDER, Samuel, 2015. *Prosperous Descent: Crisis as Opportunity in an Age of Limits.* Melbourne: Simplicity Institute Publishing. ISBN 978-0-9941606-0-7.

ALEXANDER, Samuel and Amanda McLEOD, eds., 2014. *Simple living in history: Pioneers of the deep future.* Melbourne: Simplicity Institute Publishing. ISBN 978-0-9875884-9-4.

ALEXANDER, Samuel and Simon USSHER, 2012. The Voluntary Simplicity Movement: A Multi-National Survey Analysis in Theoretical Context. *Journal of Consumer Culture*, Vol. 12, No. 1, pp. 66–86. https://doi.org/10.1177/1469540512444019

AMRINE, James W., 2002. Multiflora Rose. In: Roy VAN DRIESCHE, Bernd BLOSSEY, Mark HODDLE, Suzanne LYON and Richard REARDON, eds. *Biological Control of Invasive Plants in the Eastern United States.* Forest Health Technology Enterprise Team, pp. 265–292.

ANDER, Martin a Dušan LUŽNÝ, 2004. Lidé se bojí veřejně vyjadřovat své názory: Rozhovor s Milanem Štefancem. *Sedmá generace*, Vol. XIII, No. 5, pp. 34–37. https://doi.org/10.17104/0017-1417_2004_1_34

ARENDT, Hannah, 1998. *The Human Condition.* Chicago, London: The University of Chicago Press.

ARIÈS, Philippe, 1982. *The Hour of Our Death.* New York: Vintage Books.

ASUFU-ADJAYE, John, Linus BLOMQVIST, Stewart BRAND, Barry BROOK, Ruth DEFRIES, Erle ELLIS, Christopher FOREMAN, David KEITH, Martin LEWIS, Ted NORDHAUS, Roger JR PIELKE, Rachel PRITZKER, Joyashree ROY, Mark SAGOFF, Michael SHELLENBERGER, Robert STONE and Peter TEAGUE, 2015. *An Ecomodernist Manifesto* [online]. Accessed from: http://www.ecomodernism.org/manifesto/

AUGUSTINE, Saint, 2012. The Confessions of Saint Augustine. Grand Rapids: Christian Classics Ethereal Library.

BALHAROVÁ, Eliška, 2012. Pane, dej, aby se povedlo kroužkování. *Sedmá generace*, Vol. XXI, No. 5, pp. 12–13.

BAUMAN, Zygmunt, 1998. *Globalization: The Human Consequences.* Cambridge: Polity Press.

BECK, Ulrich, 1999. *World Risk Society.* Cambridge: Polity Press. ISBN 978-0-7456-2221-7.

BECK, Ulrich and Elisabeth BECK-GERNSHEIM, 2002. *Individualization: Institutionalized Individualism and its Social and Political Consequences.* London: SAGE. ISBN 978-1-4129-3141-0.

BĚLOHRADSKÝ, Václav, 1990. Autorita, democracies, disent. *Přítomnost*, No. 5, pp. 24–25.

BĚLOHRADSKÝ, Václav, 1991. *Myslet zeleň světa.* Prague: Mladá fronta. ISBN 80-204-0239-X.

BĚLOHRADSKÝ, Václav, 2013. *Mezi světy a mezisvěty.* 2nd revised and expanded ed. Prague: Novela Bohemica. ISBN 978-80-87683-04-0.

BERGSON, Henri, 1911. *Creative Evolution.* New York: Henry Holt and Company. https://doi.org/10.5962/bhl.title.166289

BHATTI, Mark, Andrew CHURCH, Amanda CLAREMONT and Paul STENNER, 2009. 'I Love Being in the Garden': Enchanting Encounters in Everyday Life. *Social & Cultural Geography*, Vol. 10, No. 1, pp. 61–76. https://doi.org/10.1080/14649360802553202

BINKA, Bohuslav, 2008. *Environmentální etika. Svět environmentálních souvislostí.* Brno: Masarykova univerzita. ISBN 978-80-210-4594-1.

BLÁHA, Jaromír, 2013. Wild10: Divočina, Zdraví, Láska, Bůh. *Sedmá generace*, Vol. XXII, No. 6, pp. 36–38.

BLAHUT, Petr, 2017. *Mýlit se je božské: Nová antropologie pro každého*. Brno: Doplněk.

BLAŽEK, Bohuslav and Jiří REICHEL, 1991. *Identifikace cílových skupin v oblasti ekologie a podpory zdraví*. Prague: EcoTerra3.

BLAŽEK, Jan, 2015. *The Economy of Eco-social Intentional Communities: Socio-environmental Strategies and Financial Sustainability*. Brno. PhD manuscript. Masarykova univerzita, Fakulta sociálních studií.

BLUM, Susan Debra, 2007. *Lies that bind: Chinese truth, other truths*. Lanham: Rowman & Littlefield. ISBN 978-0-7425-5404-7.

BOČEK, Jan, 2009. Ktož jsú boží zahradníci. *Sedmá generace*, Vol. XVIII, No. 2, pp. 37–38.

BONHOEFFER, Dietrich, 1959. *Prisoner for God: Letters and Papers From Prison*. New York: The Macmillan Company.

BRAND, Stewart, 2010. *Whole Earth Discipline: Why Dense Cities, Nuclear Power, Transgenic Crops, Restored Wildlands, Radical Science, and Geoengineering are Necessary*. London: Atlantic Books. ISBN 978-1-84354-816-4.

BROOKE, John Hedley, 2003. Improvable Nature? In: Willem B. DREES, ed. *Is nature ever evil? Religion, science, and value*. London: Routledge, pp. 149–169. ISBN 978-0-415-29061-6.

BROOKS, David, 2000. *Bobos in Paradise: The New Upper Class and How They Got There*. New York: Simon & Schuster.

BUBER, Martin, 1937. *I and Thou*. Edinburgh: T. & T. Clark.

CARFAGNA, Lindsey B., Emilie A. DUBOIS, Connor FITZMAURICE, Monique Y. OUIMETTE, Juliet B. SCHOR, Margaret WILLIS a Thomas LAIDLEY, 2014. An Emerging Eco-Habitus: The Reconfiguration of High Cultural Capital Practices Among Ethical Consumers. *Journal of Consumer Culture*, Vol. 14, No. 2, pp. 158–178. https://doi.org/10.1177/1469540514526227

COLUMBIA UNIVERSITY, 2016. Seeding iron in the Pacific may not pull carbon from air as thought. *Phys.org* [online]. 2. 3. 2016. Accessed from: https://phys.org/news/2016-03-seeding-iron-pacific-carbon-air.html

CONNELLY, James, 2006. The Virtues of Environmental Citizenship. In: Andrew DOBSON and Derek BELL, eds. *Environmental Citizenship*. Cambridge: MIT Press, pp. 49–73.

CRUTZEN, Paul J. and Eugene F. STOERMER, 2000. The "Anthropocene". *Global Change Newsletter*, No. 41, pp. 17–18.

CZECH CHRISTIAN ACADEMY, 1994. *Univerzum: Člověk – pastýř stvoření*. Brno: Cesta.

ČERNÝ, Václav, 1992. *První a druhý sešit o existencialismu*. Prague: Mladá fronta. ISBN 978-80-204-0337-7.

ČÍŽEK, Lukáš, Martin KONVIČKA, Jiří BENEŠ and Zdeněk FRIC, 2009. Zpráva o stavu země: Odhmyzeno. *Vesmír*, Vol. 88, No. 6, pp. 386–391.

ČRO RADIOŽURNÁL, 2015. *Twenty-minute radio programme. Interview with Minister of Transport Dan Ťok from ANO party on the Volkswagen scandal*, 24. 9. 2015.

ČSÚ, 2015. *Statistical Yearbook of the Czech Republic – 2015 – Labour market* [online]. Český statistický úřad. Accessed from: https://www.czso.cz/csu/czso/10-labour-market

DEANE-DRUMMOND, Celia, 2008. *Eco-theology*. London: Darton, Longman and Todd. ISBN 978-0-232-52616-5.

DEANE-DRUMMOND, Celia, 2014. *The Wisdom of the Liminal: Evolution and Other Animals in Human Becoming*. Grand Rapids: Eerdmans Publ. ISBN 978-0-8028-6867-1.

DEML, Jakub, 2006. Zhřešili jsme… In: *Zakletí slov II*. Prague: BB art, pp. 110–113. (Poem first published on 29. 10. 1938 in the journal *Obnova*.)

DEVALL, Bill, 1988. *Simple in Means, Rich in Ends: Practicing Deep Ecology.* Salt Lake City: Peregrine Smith Books. ISBN 978-0-87905-294-2.

DILLARD, Annie, 1974. Fecundity. In: Annie DILLARD. *Pilgrim at Tinker Creek.* New York: Harper & Row, pp. 161–183. https://doi.org/10.7591/9781501738777-046

DOBSON, Andrew, 2003. *Citizenship and the environment.* Oxford, New York: Oxford University Press. ISBN 978-0-19-925843-7.

DREES, Willem B., ed., 2003. *Is nature ever evil? Religion, science, and value.* London: Routledge. ISBN 978-0-415-29061-6.

DU CANN, Charlotte, 2012. The Dark Mountain Project: In search of a new narrative. *Energy Bulletin* [online], 28 Aug 2012. Accessed from: https://www.resilience.org/stories/2012-08-28/dark-mountain-project-search-new-narrative/

DUNLAP, Thomas R., 2004. *Faith in Nature: Environmentalism as Religious Quest.* Seattle: University of Washington Press. ISBN 978-0-295-98397-4.

DUNLAP, Thomas R., 2006. Environmentalism, a Secular Faith. *Environmental Values*, Vol. 15, No. 3, pp. 321–330. https://doi.org/10.3197/096327106778226284

DUTZIK, Tony, Jeff INGLIS and Phineas BAXANDALL, 2014. *Millennials in Motion: Changing Travel Habits of Young Americans.* U.S. PIRG Education Fund & Frontier Group.

EATON, Marie, 2012. *Environmental Trauma and Grief* [online]. Bellingham: Curriculum for the Bioregion. Fairhaven College, Western Washington University. Accessed from: http://serc.carleton.edu/bioregion/sustain_contemp_lc/essays/67207.html

EHRLICH, Paul R. and Anne H. EHRLICH, 2009. The Population Bomb Revisited. *The Electronic Journal of Sustainable Development*, Vol. 1, No. 3, pp. 63–71.

ELGIN, Duane, 1981. *Voluntary Simplicity: Toward a Way of Life That Is Outwardly Simple, Inwardly Rich.* New York: Morrow. ISBN 978-0-688-03647-8.

ELIAS, Norbert, 2000. *The civilizing process: sociogenetic and psychogenetic investigations.* Oxford, Mass.: Blackwell Publishers. ISBN 978-0-631-22160-9.

ENZENSBERGER, Hans Magnus, 1996. Reminiszenzen an den Überfluß. *Spiegel*, Vol. 51, pp. 108–118.

ERGAS, Christina, 2010. A Model of Sustainable Living: Collective Identity in an Urban Ecovillage. *Organization & Environment*, Vol. 23, No. 1, pp. 32–54. https://doi.org/10.1177/1086026609360324

EVS, 1991. *European Values Study 1990* [online]. Cologne: GESIS Data Archive. Accessed from: http://dx.doi.org/10.4232/1.10790

FABIAN, Rhonda, 2014. Shaun Chamberlin on 'Dark Optimism' and the power of grief. *Kosmos Journal* [online]. 16. 12. 2014. Accessed from: http://www.kosmosjournal.org/news/interview-shaun-chamberlin-on-dark-optimism/

FAURÉ, Marcel, 1997. *Félonie, trahison, reniements au Moyen Âge.* Proceeding of the third international colloquium of Montpellier, 24–26 November 1995. Montpellier: Université Paul Valéry.

FEDERATION OF ASIAN BISHOPS' CONFERENCES, 1993. *Love for Creation. An Asian Response to the Ecological Crisis.* Declaration of the Colloquium held in Tagatay, 31 January – 5 February 1993.

FERKANY, Matt, 2011. Mercy as an Environmental Virtue. *Environmental Values*, Vol. 20, No. 2, pp. 265–283. https://doi.org/10.3197/096327111X12997574391841

FEŘTEK, Tomáš, 2011. Je skromnost dědičná? *Ekolist.cz* [online]. 24. 8. 2011. Accessed from: http://ekolist.cz/cz/publicistika/nazory-a-komentare/tomas-fertek-je-skromnost-dedicna

FF MU, 1992. *Alternativní způsob života jako sociální realita*. Brno: Filozofická fakulta MU Sociology students' seminar work.

FLEMING, Pat and Joanna MACY, 1988. The Council of All Beings. In: John SEED, Joanna MACY, Pat FLEMING and Arne NÆSS, eds. *Thinking Like a Mountain: Towards a Council of All Beings*. Philadelphia: New Society Publishers, p. 79–90.

FOREMAN, Dave, 1992. We aren't a debating society. *Earth First! Journal*, Vol. 13, No. 2, pp. 8, 13.

FORREST, Nigel a Arnim WIEK, 2015. Success Factors and Strategies for Sustainability Transitions of Small-Scale Communities – Evidence from a Cross-Case Analysis. *Environmental Innovation and Societal Transitions*, No. 17, pp. 22–40. https://doi.org/10.1016/j.eist.2015.05.005

FOUCAULT, Michel, 1984. *Von der Freundschaft als Lebensweise: Im Gespräch*. Berlin: Merve Verlag.

FOX, Matthew, 1988. *The Coming of the Cosmic Christ: The Healing of Mother Earth and the Birth of a Global Renaissance*. San Francisco: Harper.

FRANCIS, Pope, 2015. *Laudato si': Encyclical letter on care for our common home*. Vatican City: Vatican Press.

FRASER, John, Victor PANTESCO, Karen PLEMONS, Rupanwita GUPTA and Shelley J. RANK, 2013. Sustaining the Conservationist. *Ecopsychology*, Vol. 5, No. 2, pp. 70–79. https://doi.org/10.1089/eco.2012.0076

FRASZ, Geoffrey, 2005. Benevolence as an Environmental Virtue. In: Ronald SANDLER and Philip CAFARO, eds. *Environmental virtue ethics*. Lanham: Rowman & Littlefield, pp. 121–134. ISBN 978-0-7425-3390-5.

FROMM, Erich, 1986. *You shall be as gods: A radical interpretation of the Old Testament and its tradition*. New York: Fawcett Premier.

GÁLOVÁ, Andrea, 2013. *Jak se uživit na ekovesnici? Případová studie ekovesnice Zaježka na Slovensku*. Brno. Master's thesis. Masarykova univerzita, Fakulta sociálních studií. Accessed from: http://is.muni.cz/th/385087/fss_m/

GARSD, Jasmine, 2015. Pope Francis Says Catholics Don't Need To Breed 'Like Rabbits'. *NPR* [online]. 20. 1. 2015. Accessed from: https://www.npr.org/sections/thetwo-way/2015/01/20/378559550/pope-francis-says-catholics-dont-need-to-breed-like-rabbits?t=1597839641088

GERYKOVÁ, Zuzana, 2010. *Luxusní stravování na počátku 21. století a jeho environmentální souvislosti*. Brno. Master's thesis. Masarykova univerzita, Fakulta sociálních studií. Accessed from: https://is.muni.cz/th/144355/fss_m_b1/

GORE, Al, 2006. *An inconvenient truth: the planetary emergency of global warming and what we can do about it*. New York: Rodale. ISBN 1-59486-567-1.

GRIGSBY, Mary, 2004. *Buying Time and Getting by: The Voluntary Simplicity Movement*. Albany: State University of New York Press. ISBN 978-0-7914-6000-9.

HADLER, Markus a Patrick WOHLKÖNIG, 2012. Environmental Behaviours in the Czech Republic, Austria and Germany between 1993 and 2010: Macro-Level Trends and Individual-Level Determinants Compared. *Czech Sociological Review*, Vol. 3, No. 48, pp. 467–492. https://doi.org/10.13060/00380288.2012.48.3.04

HAMILTON, Clive, 2003. *Downshifting in Britain: A Sea-change in the Pursuit of Happiness. Discussion Paper Number 58*. Deakin: Australia Institute.

HAMILTON, Clive, 2014a. Moral Collapse in a Warming World. *Ethics & International Affairs*, Vol. 28, No. 3, pp. 335–342. https://doi.org/10.1017/S0892679414000409

HAMILTON, Clive, 2014b. The Anthropocene: Too Serious for Post-Modern Games. *Clive Hamilton personal page* [online]. Accessed from: http://clivehamilton.com/the-anthropocene-too-serious-for-post-modern-games/

HAMILTON, Clive and Tim KASSER, 2009. *Psychological Adaptation to the Threats and Stresses of a Four Degree World.* "Four Degrees and Beyond" conference, Oxford University, 28–30 September 2009 [online]. Accessed from: http://clivehamilton.com/psychological-adaptation-to-the-threats-and-stresses-of-a-four-degree-world/

HANSEN, James E., 2011. *Storms of My Grandchildren: The Truth About the Coming Climate Catastrophe and Our Last Chance to Save Humanity.* New York: Bloomsbury. ISBN 978-1-60819-502-2.

HAYS, Warren S. T. a Sheila CONANT, 2007. Biology and Impacts of Pacific Island Invasive Species. 1. A Worldwide Review of Effects of the Small Indian Mongoose, Herpestes javanicus. *Pacific Science*, Vol. 61, No. 1, pp. 3–16. https://doi.org/10.1353/psc.2007.0006

HIRSCHMAN, Albert O., 2002. *Shifting Involvements: Private Interest and Public Action.* Princeton: Princeton University Press. ISBN 978-1-4008-2826-5. https://doi.org/10.1515/9781400828265

HORTON, Dave, 2003. Green distinctions: The Performance of Identity Among Environmental Activists. *The Sociological Review*, No. 51 (Supplement s2), pp. 63–77. https://doi.org/10.1111/j.1467-954X.2004.00451.x

HULL, David Lee, 1991. God of the Galápagos. *Nature*, No. 352, pp. 485–486. https://doi.org/10.1038/352485a0

INGLEHART, Ronald F., 1977. *The Silent Revolution: Changing Values and Political Styles Among Western Publics.* Princeton: Princeton University Press. ISBN 978-0-691-07585-3.

INGLEHART, Ronald F., 2008. Changing Values among Western Publics from 1970 to 2006. *West European Politics*, Vol. 31, Nos. 1–2, pp. 130–146. https://doi.org/10.1080/01402380701834747

JASPERS, Karl, 1965. *The Question of German Guilt.* New York: Fordham University Press.

JEFFERS, Robinson, 1965. *Selected Poems.* New York: Vintage Books.

JEHLIČKA, Petr, 2008. Indians of Bohemia: The spell of woodcraft on Czech society 1912–2006. In: Diethelm BLECKING and Marek WAIC, eds. *Sport-Ethnie-Nation: Zur Geschichte und Soziologie des Sports in Nationalitatenkoflikten und bei Minoritaten.* Baltmannsweiler: Schneider Verlag Hohengehren, pp. 112–130.

JEHLIČKA, Petr and Matthew KURTZ, 2013. Everyday Resistance in the Czech Landscape: The Woodcraft Culture from the Hapsburg Empire to the Communist Regime. *East European Politics and Societies*, Vol. 27, No. 2, pp. 308–332. https://doi.org/10.1177/0888325413483550

JIRÁSEK, Alois, 1992. *Old Czech legends.* London, Boston: Forest Books.

JONAS, Hans, 1987. The Concept of God after Auschwitz: A Jewish Voice. *The Journal of Religion*, Vol. 67, No. 1 (Jan 1987), pp. 1–13. https://doi.org/10.1086/487483

JONAS, Hans, 1997. *Princip odpovědnosti: Pokus o etiku pro technologickou civilizaci.* Praha: OIKOYMENH. ISBN 978-80-86005-06-5.

JONES, Olive, 2011. *Keeping It Together: A Comparative Analysis of Four Long-Established Intentional. Communities in New Zealand.* Hamilton. PhD thesis. University of Waikato. Accessed from: http://researchcommons.waikato.ac.nz/handle/10289/5962

KALA, Lukáš, 2009. *Pojetí environmentální odpovědnosti.* Brno. Master's thesis. Masarykova univerzita, Fakulta sociálních studií. Accessed from: https://is.muni.cz/th/r4u57/

KALA, Lukáš, 2014. *Environmentální aspekty* životního *způsobu českých singles.* Brno. PhD thesis. Masarykova univerzita, Fakulta sociálních studií. Accessed from: http://is.muni.cz/th/231119/fss_d/

KALA, Lukáš, Lucie GALČANOVÁ and Vojtěch PELIKÁN. 2016. Residential preferences in the context of voluntary simple lifestyles: What motivates contemporary Czech simplifiers to reside in the countryside? *Human Affairs: Postdisciplinary Humanities & Social Sciences Quarterly*, Vol. 26, No. 4, pp. 410–421. https://doi.org/10.1515/humaff-2016-0035

KALA, Lukáš, Lucie GALČANOVÁ and Vojtěch PELIKÁN. 2017. Narratives and Practices of Voluntary Simplicity in the Czech Post-Socialist Context. *Czech Sociological Review*, Vol. 53, No. 6, pp. 833–855. https://doi.org/10.13060/00380288.2017.53.6.377

KAŠPEROVÁ, Michaela, 2015. *Postenvironmentalismus v kontextu vybraných environmentálních směrů*. Brno. Bachelor's thesis. Masarykova univerzita, Fakulta sociálních studií. Accessed from: https://is.muni.cz/th/397784/fss_b/

KEVORKIAN, Kriss, 2004. *Environmental grief: Hope and healing*. Ohio. PhD thesis. Union Institute and University Cincinatti.

KINGSNORTH, Paul, 2010. Why I stopped believing in environmentalism and started the Dark Mountain Project. *The Guardian* [online]. 29. 4. 2010. Accessed from: http://www. theguardian. com/environment/2010/apr/29/environmentalism-dark-mountain-project

KINGSNORTH, Paul and Dougald HINE, 2009. *Uncivilisation: The Dark Mountain Manifesto*. The Dark Mountain Project. ISBN 978-0-9564960-6-5.

KIRBY, Andy, 2003. Redefining Social and Environmental Relations at the Ecovillage at Ithaca: A Case Study. *Journal of Environmental Psychology*, Vol. 23, No. 3, pp. 323–332. https://doi.org/10.1016/S0272-4944(03)00025-2

KLAUS, Václav, 2008. *Blue Planet in Green Shackles*. Washington: Competitive Enterprise Institute.

KLEIN, Naomi, 2014. The Change Within: The Obstacles We Face Are Not Just External. *The Nation* [online]. 21. 4. 2014. Accessed from: https://www.thenation.com/article/archive/change-within-obstacles-we-face-are-not-just-external/

KMEN, 2013. O nás: Kmen. *Kmen: Začínající komunita usilující o přirozený* život [online]. Accessed from: http://www.kmen.tode.cz/?m=201302

KOHÁK, Erazim, 2000. *The Green Halo: Bird's Eye View of Ecological Ethics*. Chicago: Open Court.

KOMÁREK, Stanislav, 2008. *Příroda a kultura: Svět jevů a svět interpretací*. Prague: Academia. ISBN 978-80-200-1582-2.

KOSOVÁ, Kateřina, 2006. *Riskantní krása květin: Environmentální a sociálně-ekonomické důsledky květinového průmyslu a informovanost spotřebitelů*. Brno. Master's thesis. Masarykova univerzita, Fakulta sociálních studií. Accessed from: https://is.muni.cz/ th/74699/fss_m/

KRAJHANZL, Jan, 2014. *Psychologie vztahu k přírodě a* životnímu *prostředí: Pět charakteristik, ve kterých se lidé liší*. Brno: Lipka – školské zařízení pro environmentální vzdělávání ve spolupráci s Masarykovou univerzitou. ISBN 978-80-87604-67-0. https://doi.org/10.5817/CZ.MUNI.M210-7063-2014

KRAUTOVÁ, Zuzana and Hana LIBROVÁ, 2009. Spotřeba domácností a proces individualizace v environmentální perspektivě. *Sociální studia*, Vol. VI, No. 3, pp. 31–55. https://doi.org/10.5817/SOC2009-3-31

KRCMAR, Marina, 2009. *Living Without the Screen: Causes and Consequences of Life without Television*. New York: Routledge. ISBN 978-0-8058-6329-1.

KROSNICK, Jon A., Allyson L. HOLBROOK, Laura LOWE and Penny S. VISSER, 2006. The Origins and Consequences of democratic citizens' Policy Agendas: A Study of Popular Concern about Global Warming. *Climatic Change*, Vol. 77, No. 1–2, pp. 7–43. https://doi.org/10.1007/s10584-006-9068-8

KUMAR, Satish, 2013. I Am An Optimist: Optimism empowers, whilst pessimism disempowers. *Resurgence*, No. 281, p. 1.

LANDY, Joshua and Michael T. SALER, eds., 2009. *The Re-enchantment of the World: Secular Magic in a Rational Age*. Standford: Stanford University Press. ISBN 978-0-8047-5299-2. https://doi.org/10.11126/stanford/9780804752992.001.0001

LÄNGLE, Alfried, 2003. Burnout – Existential Meaning and Possibilities of Prevention. *European Psychotherapy*, Vol. 4, No. 1, pp. 107–121.

LANGOVÁ, Lucie, 2012. *Houbaři a ochrana přírody*. Brno. Bachelor's thesis. Masarykova univerzita, Fakulta sociálních studií. Accessed from: https://is.muni.cz/th/329235/fss_b/

LEHEČKOVÁ, Tereza, 2016. *Kulturní chování u zvířat: rozrušení hranice lidského a mimolidského*. Brno. Bachelor's thesis. Masarykova univerzita, Fakulta sociálních studií. Accessed from: https://is.muni.cz/th/415061/fss_b/

LERTZMAN, Renée, 2015. *Environmental Melancholia: Psychoanalytic Dimensions of Engagement*. New York: Routledge. ISBN 978-0-415-72799-0.

LÉVY, Bernard-Henri, 2003. *Sartre: The Philosopher of the Twentieth Century*. Cambridge: Polity Press. https://doi.org/10.1111/j.1467-9329.2005.00282.x

LEVY, Neil, 2005. Downshifting and Meaning in Life. *Ratio*, Vol. 18, No. 2, pp. 176–189.

LEVY, Steven, 1984. *Hackers: Heroes of the Computer Revolution*. New York: Doubleday. ISBN 978-0-385-19195-1.

LIBROVÁ, Hana, 1985. Změny v sociálním hodnocení města a venkova. In: *Aktuální sociologické otázky rozvoje venkova a města*, pp. 86–96.

LIBROVÁ, Hana, 1988. *Láska ke krajině?* Brno: Blok.

LIBROVÁ, Hana, 1994. *Pestří a zelení: Kapitoly o dobrovolné skromnosti*. Brno: Veronica, Hnutí DUHA. ISBN 80-85368-18-8.

LIBROVÁ, Hana, 1996. Decentralizace osídlení – Vize a realita. Část první, Vize, postoje k venkovu a potenciální migrace v ČR. *Czech Sociological Review*, Vol. 32, No. 3, pp. 285–296. https://doi.org/10.13060/00380288.1996.32.3.04

LIBROVÁ, Hana, 1997. Decentralizace osídlení – Vize a realita. Část druhá: decentralizace v realitě České republiky. *Czech Sociological Review*, Vol. 33, No. 1, pp. 27–40. https://doi.org/10.13060/00380288.1997.33.1.06

LIBROVÁ, Hana, 1999. The Disparate Roots of Voluntary Modesty. *Environmental Values*, Vol. 8, No. 3, pp. 369–380. https://doi.org/10.3197/096327199129341879

LIBROVÁ, Hana, 2003. *Vlažní a váhaví: Kapitoly o ekologickém luxusu*. Brno: Doplněk. ISBN 80-7239-149-6.

LIBROVÁ, Hana, 2010. Individualizace v environmentální perspektivě: Sociologické rámování mění pohled a plodí otázky. *Czech Sociological Review*, Vol. 46, No. 1, pp. 125–152. https://doi.org/10.13060/00380288.2010.46.1.05

LIBROVÁ, Hana, 2013. Environmentálně orientované motivace a potenciál zklamání. *Czech Sociological Review*, Vol. 49, No. 1, pp. 53–74. https://doi.org/10.13060/00380288.2013.49.1.03

LIBROVÁ, Hana, 2014a. Environmentální žal. *Vesmír*, Vol. 93, No. 4, pp. 238–241.

LIBROVÁ, Hana, 2014b. Otec Kondelík. *Respekt*, Vol. XXV, No. 39, pp. 70–71.

LOMBORG, Bjørn, 2001. *The Skeptical Environmentalist: Measuring the Real State of the World*. Cambridge: Cambridge University Press. https://doi.org/10.1017/CBO9781139626378

LOQUENZ, Jan a Martin ŠIMON, 2013. Amenitní migrace. *Geografické rozhledy*, No. 2, pp. 26–17.

LORENZ, Konrad, 1966. *On Aggression*. London: Methuen.

LORENZ, Konrad, 1974. *Civilized Man's Eight Deadly Sins*. New York: Harcourt Brace Jovanovich

LORENZEN, Janet A., 2012. Green and Smart: The Co-Construction of Users and Technology. *Human Ecology Review*, Vol. 19, No. 1, pp. 25–36.

LOUV, Richard, 2012. *The Nature Principle: Reconnecting with Life in a Virtual Age*. Chapel Hill: Algonquin Books. ISBN 978-1-61620-141-8.

LOVELOCK, James, 1991. Planetary Medicine. *Resurgence*, No. 148, pp. 38–45. https://doi.org/10.2307/1088552

LOVELOCK, James. 2000. *Gaia: A New Look at Life on Earth*. Oxford: Oxford University Press.

LOVELOCK, James. 2006. *The Revenge of Gaia: Why the Earth is Fighting Back and How We Can Still Save Humanity*. London: Allen Lane.LOVELOCK, James, 2014. *A Rough Ride to the Future*. London: Allen Lane. ISBN 978-0-241-00476-0.

LUKAŠÍKOVÁ, Monika, 2009. *Secondhandy – environmentálně uvědomělé nakupování?* Brno. Bachelor's thesis. Masarykova univerzita, Fakulta sociálních studií. Accessed from: https://is.muni.cz/th/219925/fss_b/

LUŽNÝ, Dušan, 2000. *Zelení bódhisattvové: sociálně a ekologicky angažovaný buddhismus*. Brno: Masarykova univerzita. ISBN 978-80-210-2464-9.

LYNAS, Mark, 2011. *The God Species: Saving the Planet in the Age of Humans*. Washington D.C.: National Geographic. ISBN 978-1-4262-0891-1.

MacINTYRE, Alasdair, 1981. *After Virtue: A Study in Moral Theory*. Notre Dame: University of Notre Dame Press.

MACY, Joanna and Molly Young BROWN, 1998. *Coming Back to Life: Practices to Reconnect Our Lives, Our World*. Gabriola Island, Stony Creek: New Society Publishers. ISBN 978-0-86571-391-8.

MACY, Joanna, 1988. Bestiary. In: John SEED, Joanna MACY, Pat FLEMING a Arne NÆSS, eds. *Thinking Like a Mountain: Towards a Council of All Beings*. Philadelphia: New Society Publishers, pp. 74–78.

MAFFESOLI, Michel, 1997. *Du nomadisme. Vagabondages initiatiques*. Paris: Livre de Poche.

MANDER, Jerry, 1991. *In the Absence of the Sacred: The Failure of Technology and the Survival of the Indian Nations*. San Francisco: Sierra Club Books. ISBN 978-0-87156-509-9.

MANIATES, Michael, 2001. Individualization: Plant a Tree, Buy a Bike, Save the World? *Global Environmental Politics*, Vol. 1, No. 3, pp. 31–52. https://doi.org/10.1162/152638001316881395

MANIATES, Michael, 2002. In Search of Consumptive Resistance: The Voluntary Simplicity Movement. In: Thomas PRINCEN, Michael MANIATES a Ken CONCA, eds. *Confronting Consumption*. Cambridge: MIT Press, pp. 199–235. ISBN 978-0-262-66128-7.

MARCEL, Gabriel, 1951. *Homo Viator: Introduction to a Metaphysic of Hope*. Chicago: Henry Regnery Company.

MARCEL, Gabriel, 1969. *The Philosophy of Existence*. New York: Books For Libraries Press.

MARIETTE, Mylene M. and Katherine L. BUCHANAN, 2016. Prenatal acoustic communication programs offspring for high posthatching temperatures in a songbird. *Science*, Vol. 353, No. 6301, pp. 812–814. https://doi.org/10.1126/science.aaf7049

MARRIS, Emma, 2013. *Rambunctious Garden: Saving Nature in a Post-wild World*. New York: Bloomsbury. ISBN 978-1-60819-454-4.

MARTINEZ, Demetria, 2010. Environmental Grief so Great. *National Catholic Reporter* [online]. 6. 7. 2010. Accessed from: https://www.ncronline.org/blogs/ncr-today/environmental-grief-so-great

MARTÍNEZ-ALIER, Joan, 1995. The Environment as a Luxury Good or 'Too Poor to Be Green'? *Ecological Economics*, Vol. 13, No. 1, pp. 1–10. https://doi.org/10.1016/0921-8009(94)00062-Z

McKIBBEN, Bill, 2011. *Eaarth: Making a life on a tough new planet*. New York: St. Martin's Griffin.

McLELLAN, Richard, Leena IYENGAR, Barney JEFFRIES and Natasja OERLEMANS, eds., 2014. *Living Planet Report 2014: Species and Spaces, People and Places* [online]. Gland, Switzerland: WWF International. Accessed from: http://www.worldwildlife.org/pages/living-planet-report-2014

MEADOWS, Donella H., Denis L. MEADOWS, Jørgen RANDERS and William W. BEHRENS, eds., 1972. *The Limits to Growth: A Report for the Club of Rome's Project on the Predicament of Mankind.* New York: Universe Books. ISBN 0-87663-165-0. https://doi.org/10.1349/ddlp.1

MERTON, Robert King, 1968. *Social Theory and Social Structure.* New York: The Free Press.

MISÍKOVÁ, Marta, 1994. *Sociální a demografická charakteristika* účastníků *akce FALCO.* Brno. Bachelor's thesis. Masarykova univerzita, Filosofická fakulta.

MOLDAN, Bedřich, 2009. *Podmaněná planeta.* Prague: Karolinum. ISBN 978-80-246-1580-6.

MOLLISON, Bill, 1988. *Permaculture: A Designer's Manual.* Tyalgum: Tagari Publications. ISBN 978-0-908228-01-0.

MOLTMANN, Jürgen, 1993. *God in Creation: A New Theology of Creation and the Spirit of God.* Minneapolis: Fortress Press.

MONBIOT, George, 2014. *Feral: Rewilding the Land, the Sea, and Human Life.* Chicago: The University of Chicago Press. ISBN 978-0-226-20555-7. https://doi.org/10.7208/chicago/9780226205694.001.0001

MONBIOT, George, 2015a. We're Not as Selfish as We Think We Are. Here's the Proof. *The Guardian* [online]. 14. 10. 2015. Accessed from: http://www.theguardian.com/commentisfree/2015/oct/14/selfish-proof-ego-humans-inherently-good

MONBIOT, George, 2015b. Indonesia is burning. So why is the world looking away? *The Guardian* [online]. 30. 10. 2015. Accessed from: https://www.theguardian.com/commentisfree/2015/oct/30/indonesia-fires-disaster-21st-century-world-media

MRÁZEK, Jiří, 2012. *Zjevení Janovo.* Prague: Centrum biblických studií AVČR a UK, Česká biblická společnost. ISBN 978-80-85810-99-8.

MUSIL, Libor, 1992. *Možnosti a meze konsensu o otázce Novomlýnských nádrží – Případová studie konfliktu.* Brno-Dražovice: EGE Buzek.

NÆSS, Arne, 1973. The Shallow and the Deep, Long-Range Ecology Movement. *Inquiry*, Vol. 16, Nos. 1–4, pp. 95–100.

NÆSS, Arne, 1990a. 'Man Apart' and Deep Ecology: A Reply to Reed. *Environmental Ethics*, Vol. 12, No. 2, pp. 185–192. https://doi.org/10.5840/enviroethics199012224

NÆSS, Arne, 1990b. *Sustainability: The integral approach.* In: Proceedings of the International Conference of Genetic Resources for Sustainable Development, Røros, Norway, Sep 10–14.

NĚMEC, Václav, 2010. Filosof jako zoon antipolitikon: Vztah filosofie a politiky v antice podle H. Arendtové. *Aithér*, Vol. 2, No. 3, pp. 164–180. https://doi.org/10.5507/aither.2010.012

NEUBAUER, Zdeněk, 2002. *Biomoc.* Prague: Malvern. ISBN 978-80-902628-6-7.

NOLAN, Christopher, 2014. *Interstellar.* Film. USA, Great Britain. 169 min. https://doi.org/10.5040/9780571343089-div-00000012

O'NEILL, John, Alan HOLLAND and Andrew LIGHT, 2008. *Environmental Values.* London, New York: Routledge. ISBN 978-0-415-14508-4.

ODUM, Eugene Pleasants, 1953. *Fundamentals of Ecology.* Philadelphia: W. B. Saunders Company.

ORR, David W., 1994. *Earth in Mind: On Education, Environment, and the Human Prospect.* Washington D.C.: Island Press. ISBN 978-1-55963-495-3.

ORR, David W., 1996. Slow Knowledge. *Resurgence*, No. 179, pp. 30–32.

PAUSEWANG, Gudrun, 1980. *Rosinkawiese. Alternatives Leben vor 50 Jahren*. Ravensburg: Maier. ISBN 978-3-473-35055-1.

PAVALACHE-ILIE, Mariela and Ecaterina Maria UNIANU, 2012. Locus of Control and the Pro-environmental Attitudes. *Procedia – Social and Behavioral Sciences*, No. 33, pp. 198–202. https://doi.org/10.1016/j.sbspro.2012.01.111

PECHOVÁ, Martina, 2001. Max Weber – Myslitel a politik. *Czech Sociological Review*, Vol. 37, No. 4, pp. 463–471. https://doi.org/10.13060/00380288.2001.37.4.05

PELIKÁN, Vojtěch and Hana LIBROVÁ, 2015. Motivation for Environmental Direct Action in the Czech Republic: The Case of the 2011 Blockade at the Šumava National Park. *Sociální studia*, Vol. XII, No. 3, pp. 27–52. https://doi.org/10.5817/SOC2015-3-27

PEŘINA, Vlastimil, 2015. Tak tedy, co se osvědčilo, na co si dát pozor, na co je třeba myslet. *Ptačí svět*, No. 3, p. 25.

PETRŮ, Marek, 2005. *Možnosti transgrese: Je třeba vylepšovat* člověka? Prague: Triton. ISBN 978-80-7254-610-7.

PETRUSEK, Miloslav, 1986. *Transgrese jako* životní *způsob*. Brno: Sympozium Socialistický způsob života jako sociální realita III, pp. 92–101.

PLESNÍK, Jan and Václav PETŘÍČEK, 2012. Významné krajinné prvky a ekologická stabilita. *Ochrana přírody*, zvláštní číslo, pp. 41–44.

PODHORNÁ-POLICKÁ, Alena, 2011. *Neologie a dynamika* šíření *identitárních neologismů mezi mládeží*. Prague: Univerzita Karlova, Filozofická fakulta. Přednáška proslovená v Kruhu moderních filologů.

POKORNÝ, Petr, ed., 2008. *Něco překrásného se končí: Kolapsy v přírodě a společnosti*. Prague: Dokořán. ISBN 978-80-7363-197-0.

POLLAN, Michael, 1991. *Second Nature: A Gardener's Education*. New York: Grove Press. ISBN 978-0-8021-4011-1.

POLLARD, Dave, 2012. Giving Up on Environmentalism. Blog *How to Save the World* [online]. 12. 4. 2012. Accessed from: http://howtosavetheworld.ca/2012/04/10/giving-up-on-environmentalism/

POPELKOVÁ, Monika, 2013. *Luxus v klášteře: Environmentální souvislosti* životního *způsobu klarisek*. Brno. Master's thesis. Masarykova univerzita, Fakulta sociálních studií. Accessed from: https://is.muni.cz/th/144493/fss_m_a2/

PORQUET, Jean-Luc, 2016. Rien sur le pare-brise. *Le Canard enchaîné*. 14. 12. 2016.

PORRITT, Jonathon, 2007. *Capitalism As If The World Matters*. London, Sterling: Earthscan. ISBN 978-1-84407-193-7.

PSO, 2010. Sloboda, ohľaduplnosť, ekologická zodpovednosť. *Pôdne Spoločenstvo Olšinka* [online]. Accessed from: http://olsinka.eu/

RABUŠIC, Ladislav, 2001. Value change and demographic behaviour in the Czech Republic. *Czech Sociological Review*, Vol. 9, No. 1, pp. 99–122. https://doi.org/10.13060/00380288.2001.37.11.15

RAY, Paul H. and Sherry Ruth ANDERSON, 2000. *The Cultural Creatives: How 50 Million People are Changing the World*. New York: Three Rivers Press. ISBN 978-0-609-80845-0.

REID, Siân Lee and Shelley Tsivia RABINOVITCH, 2008. Witches, Wiccans, and Neo-Pagans: A Review of Current Academic Treatments of Neo-Paganism. In: James R. LEWIS, ed. *The Oxford handbook of new religious movements*. Oxford, New York: Oxford University Press, pp. 514–533. ISBN 978-0-19-536964-9.

REID, Walter V., Harold A. MOONEY and Angela CROPPER, 2005. *Ecosystems and Human Well-being: Synthesis*. Washington, D.C.: Island Press.

RIDLEY, Matt, 1996. *The Origins of Virtue*. London: Penguin Books.

ROBERTS, David, 2007. Czech pres. isn't on board with the climate change thing. *Grist.org* [online], 14. 2. 2007. Accessed from: https://grist.org/article/czech-bait/

ROBOVSKÝ, Jan, 2012. Naděje pro zvířata a jejich prostředí, povedená a potřebná kniha. *Vesmír*, Vol. 91, No. 1, pp. 54–55.

RODIN, Judith and Ellen J. LANGER, 1977. Long-Term Effects of a Control-Relevant Intervention With the Institutionalized Aged. *Journal of Personality and Social Psychology*, Vol. 35, No. 12, pp. 897–902. https://doi.org/10.1037/0022-3514.35.12.897

ROTTER, Julian B., 1954. *Social Learning and Clinical Psychology*. New York: Prentice-Hall. https://doi.org/10.1037/10788-000

ROZBOŘIL, Blahoslav, 1992. *Přítel Martin – vegetarián, který neodmítne* řízek. Brno. Seminar work. Filozofická fakulta MU.

SALE, Kirkpatrick, 1985. *Dwellers in the Land: The Bioregional Vision*. Georgia: University of Georgia Press. ISBN 978-0-8203-2205-6.

SANDLER, Ronald a Philip CAFARO, eds., 2005. *Environmental Virtue Ethics*. Lanham: Rowman & Littlefield. ISBN 978-0-7425-3390-5.

SANIDES-KOHLRAUSCH, Claudia, 2003. The Lisbon Earthquake. In: Willem B. DREES, ed. *Is Nature Ever Evil? Religion, Science, and Value*. London: Routledge, pp. 106–119. ISBN 978-0-415-29061-6.

SARX, 2015. *Christian Animal Welfare* [online]. Accessed from: http://sarx.org.uk

SAX, Boria, 2000. *Animals in the Third Reich: Pets, Scapegoats, and the Holocaust*. New York, London: Continuum.

SCRUTON, Roger, 2006. Conserving Nature. In: Roger SCRUTON. *A Political Philosophy: Arguments for Conservatism*. New York: Continuum, pp. 32–46.

SCRUTON, Roger, 2010. *The Uses of Pessimism and the Danger of False Hope*. New York: Atlantic Books. ISBN 978-1-84887-881-5.

SEED, John, 1988. Introduction: "To Hear Within Ourselves the Sound of the Earth Crying". In: John SEED, Joanna MACY, Pat FLEMING and Arne NÆSS, eds. *Thinking Like a Mountain: Towards a Council of All Beings*. Philadelphia: New Society Publishers, pp. 5–18.

SEGAL, Lynne, 1990. *Slow Motion – Changing Masculinities, Changing Men*. London: Virago.

SHELLENBERGER, Michael and Ted NORDHAUS, 2004. *The Death of Environmentalism: Global Warming Politics in a Post-Environmental World* [online]. Oakland: Breakthrough Institute. Accessed from: http://www.thebreakthrough.org/images/Death_of_Environmentalism.pdf

SCHERER, Georg, 2005. *Tomáš Akvinský s Georgem Schererem*. Kostelní Vydří: Karmelitánské nakladatelství. ISBN 80-7192-947-6.

SCHMIDT, Karl, 1993. On Economization and Ecologization as Civilizing Processes. *Environmental Values*, Vol. 2, No. 1, pp. 33–46. https://doi.org/10.3197/096327193776679963

SCHOR, Juliet B., 1999. *The Overspent American: Why We Want What We Don't Need*. New York: Harper Collins. ISBN 978-0-06-097758-0.

SCHOR, Juliet B., Connor FITZMAURICE, Lindsey B. CARFAGNA and Will ATTWOOD-CHARLES, 2016. Paradoxes of Openness and Distinction in the Sharing economy. *Poetics*, No. 54, pp. 66–81. https://doi.org/10.1016/j.poetic.2015.11.001

SCHUMACHER, Ernst F., 1973. *Small Is Beautiful: A Study of Economics As If People Mattered*. New York: Harper & Row.

SCHÜTZE, Christian, 1989. *Das Grundgesetz vom Niedergang: Ruiniert Arbeit die Welt?* München: Carl Hanser Verlag.

SCHWEITZER, Albert, 1990. *Out of My Life and Thought: An Autobiography*. New York: John MacRae Books.

SKALÍK, Jan and Matěj BAJGAR, 2012. *Nalezeni Na Ztraceném*. Documentary film. CR. 42 min.

SKOVAJSA, Marek, 2012. Nadvláda účelové racionality? K interpretacím Weberovy typologie jednání. *Sociológia*, Vol. 44, No. 5, pp. 579–601.

SMELLIE, William, 1836. *The Philosophy of Natural History*. Boston: Hilliard Gray.

SMITH, Adam. 1976. *An inquiry into the nature and causes of the wealth of nations*. The Glasgow Edition of the Works and Correspondence of Adam Smith, Vol. 2. Oxford: Oxford University Press.

SMITH, Daniel, 2014. It's the end of the world as we know it... and he feels fine. *The New York Times Magazine* [online], 7. 4. 2014. Accessed from: http://www.nytimes.com/ 2014/04/20/ magazine/its-the-end-of-the-world-as-we-know-it-and-he-feels-fine.html

SNYDER, Gary, 1969. *Four Changes*. Santa Barbara: The Unicorn Bookshop. https://doi.org/10.7901/2169-3358-1969-1-309

SOPER, Kate, 2008. Alternative hedonism, cultural theory and the role of aesthetic revisioning. *Cultural Studies*, Vol. 22, No. 5, pp. 567–587. https://doi.org/10.1080/09502380802245829

SOUKUP, Daniel, 2013. *„Cikáni" a česká vesnice: konstrukty cizosti v literatuře 19. století*. Prague: Nakladatelství Lidové noviny.

SPASH, Clive L., 2015. The Dying Planet Index: Life, Death and Man's Domination of Nature. *Environmental Values*, Vol. 24, No. 1, pp. 1–7. https://doi.org/10.3197/09632711 5X14183182353700

SPURNÝ, Miloš, 2007. *Sbohem, staré řeky / Good-Bye, Old Rivers*. Brno: FOTEP. ISBN 978-80-86871-07-3.

STEELE, Michael A., Sylvia L. HALKIN, Peter D. SMALLWOOD, Thomas J. MCKENNA, Katerina MITSOPOULOS and Matthew BEAM, 2008. Cache protection strategies of a scatterhoarding rodent: do tree squirrels engage in behavioural deception? *Animal Behaviour*, Vol. 75, No. 2, pp. 705–714. https://doi.org/10.1016/j.anbehav.2007.07.026

STIBOR, Vladimír *et al.*, 2016. Řezbáři *stínů: Almanach* české *poezie 2016*. Prague: Milan Hodek. ISBN 978-80-87688-53-3.

STIBRAL, Karel, 2005. *Proč je příroda krásná?: estetické vnímání přírody v novověku*. Prague: Dokořán. ISBN 80-7363-008-7.

STORCH, David, 2008. Katastrofy v přírodě: statistika, dynamika a teorie. In: Petr POKORNÝ and Miroslav BÁRTA, eds. *Něco překrásného se končí: kolapsy v přírodě a společnosti*. Prague: Dokořán, pp. 31–36. ISBN 978-80-7363-197-0.

SVOBODOVÁ, Renata, 2016. Mizí krajinná mozaika, divočina to nezachrání: Rozhovor s Davidem Storchem. *Sedmá generace*, Vol. XXV, No. 2, pp. 55–59. https://doi.org/10.1007/978-3-319-22596-8_5

ŠEFARA, Denis, Marek FRANĚK and Václav ZUBR, 2015. Socio-psychological Factors that Influence Car Preference in Undergraduate Students: The Case of the Czech Republic. *Technological and Economic Development of Economy*, Vol. 21, No. 4, pp. 643–659. https://doi.org/10.3846/202949 13.2015.1055617

ŠMÍDOVÁ, Iva, 2009. Changing Czech Masculinities? Beyond Environment And Children Friendly Men. In: Elżbieta H. OLEKSY, ed. *Intimate Citizenships: Gender, Sexualities, Politics*. New York: Routledge, pp. 193–206. ISBN 978-0-415-99076-9.

ŠTĚPÁNEK, Radek, 2018. *Hic Sunt Homines*. Brno: Masarykova univerzita. ISBN: 978-80-210-8882-5

ŠTĚPÁNKOVÁ, Kateřina, 2016. *Dárky jako významná složka spotřeby*. Brno. Master's thesis. Masarykova univerzita, Fakulta sociálních studií. Accessed from: https:// is.muni.cz/auth/ th/432072/fss_m/

ŠŮLOVÁ, Karolína, 2012. Rozhovor s Davidem Storchem. *Ochrana přírody*, No. 5, pp. 32–33.

TEIGE, Karel, 1947. Předmluva o architektuře a přírodě. In: Ladislav ŽÁK. *Obytná krajina*. Prague: Spolek výtvarných umělců Mánes, pp. 7–21.

TNS, 2016. *Connected Life : Czech Republic Report*. Unpublished research.

TREANOR, Brian, 2008. Narrative Environmental Virtue Ethics: Phronesis without a Phronimos. *Environmental Ethics*, Vol. 30, No. 4, pp. 361–379. https://doi.org/10.5840/ enviroethics200830440

TURNER, Fred, 2006. *From Counterculture to Cyberculture: Stewart Brand, the Whole Earth network, and the Rise of Digital Utopianism*. Chicago, London: University of Chicago Press. ISBN 978-0-226-81742-2. https://doi.org/10.7208/chicago/9780226817439.001.0001

ULČÁK, Zbyněk, 2015. *Život je černý i bílý*. Drnovice: Občanské sdružení Drnka. ISBN 978-80-904591-4-4.

VÁNĚ, Jan, 2012. *Komunita jako nová naděje? Náboženské (ne)institucionalizované komunity z pohledu sociologie náboženství*. Plzeň: Západočeská univerzita v Plzni. ISBN 978-80-261-0085-0.

VERBEEK, Bernard, 1998. *Die Anthropologie der Umweltzerstörung: Die Evolution und der Schatten der Zukunft*. Darmstadt: Primus.

VOJTÍŠEK, Zdeněk, 2007. Novopohanství a jeho přítomnost v české společnosti. *Teologická revue* [online]. Accessed from: http://www.zvojtisek.cz/dokumenty/clanky/novo-pohanstvi_pro_ teologickou_revue.pdf

WALLETSCHEK, Hartwig a Jochen GRAW, 1990. *Ökolexikon: Stichworte und Zusammenhänge*. München: Verlage C.H. Beck und Biederstein.

WALTHER, Carol S. and Jennifer A. SANDLIN, 2013. Green Capital and Social Reproduction within Families Practising Voluntary Simplicity in the US. *International Journal of Consumer Studies*, Vol. 37, No. 1, pp. 36–45. https://doi.org/10.1111/j.1470-6431.2011.01050.x

WEBER, Max, 1922a. *Gesammelte Aufsätze zur Religionssoziologie I*. Tübingen: J.C.B. Mohr (Paul Siebeck).

WEBER, Max, 1922b. *Wirtschaft und Gesellschaft*. Tübingen: J.C.B. Mohr (Paul Siebeck).

WEBER, Max, 1946. *Politics as a Vocation*. New York: Oxford University Press.

WEINTROBE, Sally, ed., 2012. *Engaging with Climate Change: Psychoanalytic and Interdisciplinary Perspectives*. London: Routledge. ISBN 978-0-415-66760-9. https://doi.org/10.4324/9780203094402

WELCH, Bryan, 2011. What a Great Time to Be Alive! *Mother Earth News* [online]. 11. 10. 2011. Accessed from: http://www.motherearthnews.com/nature-and-environment/ what-a-great-time-to-be-alive.aspx

WERBACH, Adam, 2010. Debating 'What the Green Movement Got Wrong'. *The Atlantic* [online]. 5. 11. 2010. Accessed from: http://www.theatlantic.com/entertainment/ archive/2010/11/ debating-what-the-green-movement-got-wrong/66162/

WHITE, Lynn, 1967. The Historical Roots of Our Ecologic Crisis. *Science*, Vol. 155, No. 3767, pp. 1203–1207. https://doi.org/10.1126/science.155.3767.1203

WHITMARSH, Lorraine, 2011. Scepticism and Uncertainty About Climate Change: Dimensions, Determinants and Change Over Time. *Global Environmental Change*, Vol. 21, No. 2, pp. 690–700. https://doi.org/10.1016/j.gloenvcha.2011.01.016

WILSON, Edward O., 1984. *Biophilia*. Cambridge: Harvard University Press. ISBN 978-0-674-07441-5.

WILSON, Edward O., 1992. *The Diversity of Life*. Cambridge: Harvard University Press.

WILSON, Edward O., 2006. *The Creation: An Appeal to Save Life on Earth*. New York, London: W. W. Norton & Company. ISBN 978-0-393-06217-5.

WINDLE, Phyllis, 1995. The Ecology of Grief. *BioScience*, No. 42, pp. 363–366. https://doi.org/10.2307/1311783

WINTER, Deborah Du Nann and Susan M. KOGER, 2010. *The Psychology of Environmental Problems: Psychology for Sustainability*. New York: Psychology Press.

WORSTER, Donald, 1985. *Nature's Economy: A History of Ecological Ideas*. Cambridge, New York: Cambridge University Press. ISBN 0-521-31870-X.

WUERTHNER, George, Eileen CRIST and Tom BUTLER, eds., 2014. *Keeping the Wild: Against the Domestication of Earth*. Washington, D.C.: Island Press. ISBN 978-1-61091-558-8. https://doi.org/10.5822/978-1-61091-559-5

WUKETITS, Franz M., 1998. *Naturkatastrophe Mensch. Evolution ohne Fortschritt*. Düsseldorf: Patmos .

WUTHNOW, Robert, 1996. *Poor Richard's Principle: Recovering the American Dream through the Moral Dimension of Work, Business, and Money*. Princeton: Princeton University Press. ISBN 978-1-4008-1392-6.

WWF, 2016. *Living Planet Report 2016. Risk and resilience in a new era* [online]. Gland, Switzerland: WWF International. Accessed from: http://wwf.panda.org/about_our_earth/all_publications/lpr_2016/

YALOM, Victor, 2010. Kenneth Doka on Grief Counseling and Psychotherapy. *Psychotherapy.net* [online]. Accessed from: https://www.psychotherapy.net/interview/grief-counseling-doka

ZAMWEL, Einat, Orna SASSON-LEVY and Guy BEN-PORAT, 2014. Voluntary simplifiers as political consumers: Individuals practicing politics through reduced consumption. *Journal of Consumer Culture*, Vol. 14, No. 2, pp. 199–217. https://doi.org/10.1177/1469540514526277

ZBÍRAL, David, 2007. New Age a novopohanství. *David Zbíral – religionistika* [online]. Accessed from: http://www.david-zbiral.cz/NewAge.htm

ZWACH, Ivan, 2009. *Obojživelníci a plazi* České republiky: encyklopedie všech druhů, určovací klíč. Prague: Grada. ISBN 978-80-247-2509-3.

Biblical quotations are from the New International Version (online)

MASARYK
UNIVERSITY
MONOGRAPHS
Vol. 1

HANA LIBROVÁ

and Vojtěch Pelikán – Lucie Galčanová Batista – Lukáš Kala

THE FAITHFUL
AND
THE REASONABLE

CHAPTERS ON ECOLOGICAL FOOLISHNESS

Illustrations by Bohdan Lacina
Translated by Graeme Dibble
Edited by Vojtěch Pelikán and Alena Mizerová
Layout and typesetting by Eva Lufferová and Jan Búřil
Published by Masaryk University Press
Printed by PowerPrint

Published as the 1st volume in the Series "Masaryk University Monographs"
supported by Scientia est Potentia fund
First Edition
2021

ISBN 978-80-210-9991-3

https://doi.org/10.5817/CZ.MUNI.M210-9992-2021

www.ingramcontent.com/pod-product-compliance
Lightning Source LLC
Chambersburg PA
CBHW041731100726
47973CB00011B/179